NO GRID Survival Projects

Transformative Projects for Living Without Limits

Disclaimer

This book is designed only to provide information about do-it-yourself projects for survival and self-sufficiency. **No Grid Survival Projects** is sold with the knowledge that the publisher, editor, and authors do not offer any legal or other professional advice regarding this or any other subject. In the case of a need for any such expertise, you should always consult the appropriate professional, whether that is an electrician, plumber, or other.

This book does not contain all the information available on the subject of these DIY projects.

It has not been created to be specific to any individual's or organization's situation or needs. While the authors, editor, and publisher have made every effort to make the knowledge inside this book as accurate as possible there may still exist typographical and/or content errors that have made it through. Therefore, this book should serve only as a general guide and not as the ultimate source of information on the subject of DIY projects.

The foods described in this book do not comply with FDA, USDA, or FSIS regulations or local health codes. Dehydrating meat products does not reduce the health risks associated with meat contaminated with Salmonella and/or E. coli O157H7. The authors, editor, and publisher cannot and will not guarantee the shelf-life of any of the food products described in this book. Eating the foods will be done at your own risk and it will be your responsibility to test them for microbes, bacteria, viruses, or germs before consuming them.

The authors, editor, and publisher shall have no liability or responsibility to any person or entity regarding any losses or damage incurred, or alleged to have incurred, directly or indirectly, by the information contained in this book. The instructions provided have not been reviewed, tested, or approved by any official testing body or government agency; you agree that by using them you waive any right of litigation.

The authors and editor of this book make no warranty of any kind, expressed or implied, regarding the safety of the final products or of the methods used. Building and operating any of the projects described in this book will be done at your own risk. This includes but is certainly not limited to any physical injury which may occur to yourself or others by incorrectly or unsafely using the knowledge inside this book.
By reading past this point you hereby agree to be bound by this disclaimer, or you may return this book within the guarantee time period for a full refund.

Table of Contents

DIY Sink To Save Water

MATERIAL COST  99.00 EASY  DIFFICULTY 30 MINUTES 

Your off grid water supply is precious. The amount of water you have on hand depends on what kind of water catchment system you have and the type of area you live in. No matter how much water you store, conservation of water is always important. Off grid water might come from the rain, a creek, or even a well. If you are using toilets that flush rather then composting toilets, then you need water to flush them. Modern toilets use around 2 gallons of water every time you flush, while older toilets can use as much as 7 gallons per flush!

This simple DIY project will allow you to turn your toilet into a sink on every flush. By running the water to fill your toilet, both into the bowl and up through this simple faucet, you will be able to wash your hands with the same water you use to fill your toilet bowl.

There will be no water lost handwashing after using the bathroom. Every drop of water used will pour over your hands and right backdown into the reservoir that will flush your toilet the next time. Even better, the soap you use will also be rinsed off into that reservoir and when you flush your toilet next the bowl will be flushed with soapy water. This helps keep your toilet clean, too!

Materials Needed

In order to put together this project, you will need:

- Sink Twice Sink Fixture $99.00
- Included Tubes and Faucet Attachment
- Extender for Better Fit

www.amazon.com/Twice-gloss-expander-larger-measured/dp/B08DJC9WN1/ref

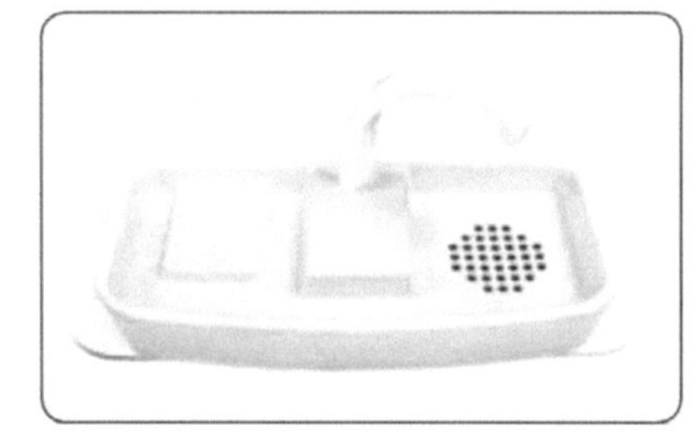

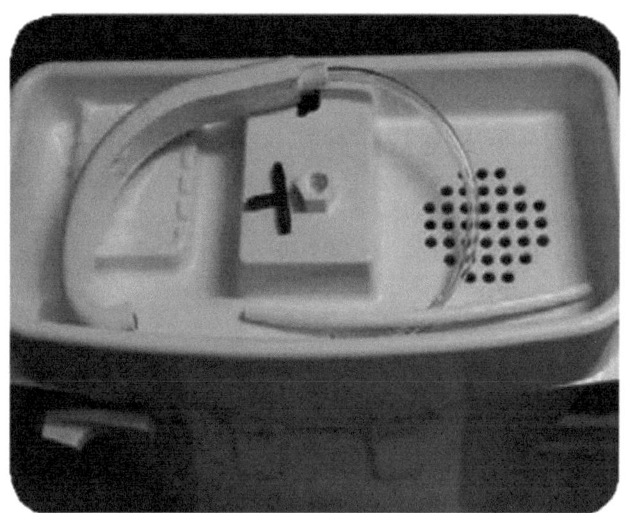

The Building Process

1. Measure your toilet before ordering your Sink Twice fixture. There are a couple of different models depending on the size of your toilet. This is an important step because you could waste a lot of money if you buy the wrong size and it doesn't fit.
2. Once you have received your Sink Twice, you are going to remove the porcelain top from your toilet and check the fit.
3. If you bought the extenders or bought a unit that includes them, you will be able to install those now too. That is an easy process of just slipping both sides inside the unit.
4. Now that you know the unit fits properly, you can begin the work to reroute the water that fills your toilet after every flush to run up through the Sink Twice, too.
5. Start by assembling the faucet by feeding the faucet tube through the hole in the sink basin.

6. Slide the included hex nut over the faucet tube on the underside of the sink and screw it, tightly, into the facet tube.
7. At this point we are going to move into the toilet tank and find our refill tube, refill valve, and overflow tube.
8. We are going to take our fill cycle diverter or this black T shaped part and insert a clear flow modifier. This will divert the water to your sink faucet. The flow modifier should be in the downward position and the short end of the fill cycle diverter should be pointed towards your refill tube.
9. Attach the refill tube to the short end of the diverter and then attach the sink faucet to the tube to the diverter end that is pointing up. You may need to trim the faucet tube. It should be straight. Any crimping will affect water pressure to the sink.
10. Be sure the fill cycle diverter is pointing its lower prong into the overflow tube. You could even attach a small length of tube to this part of the diverter and feed it into the overflow tube for stability of the entire setup.
11. Check all your connections and then place the sink basin over the toilet tank.
12. Check that the unit is levelled. This is required for the water to properly drain back into the tank.
13. This is where the expanders can come in handy. In some models of the Sink Twice these are not included, but I did find one where they were included.
14. At this point you should be ready to give the toilet a flush and test out the whole system.
15. It is recommended that you use sea mineral soap or unscented foam soap as other may have an affect on the toilet bowl over time.
16. Your DIY water saving sink is now installed and you should give it a try.

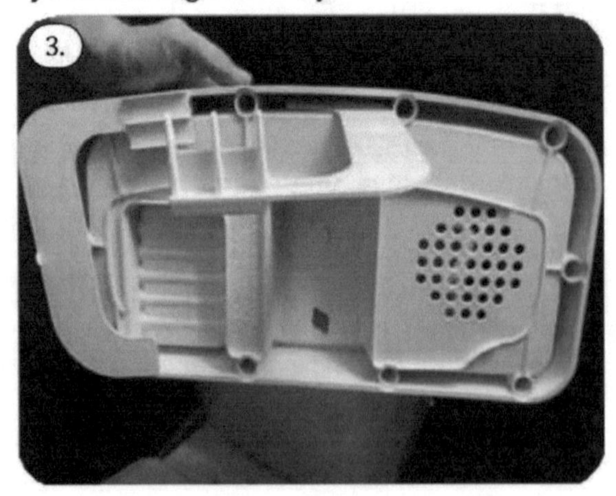

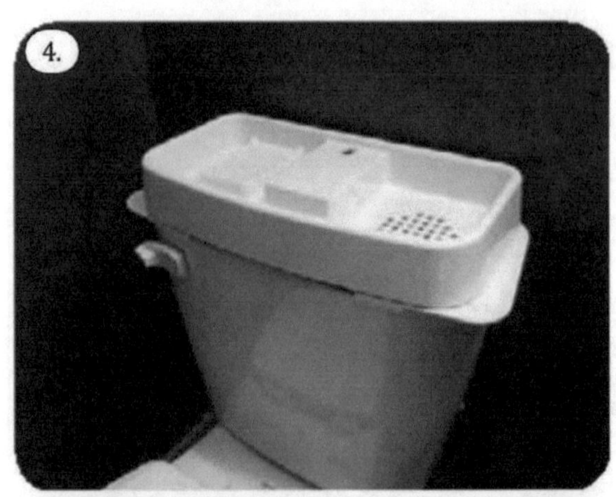

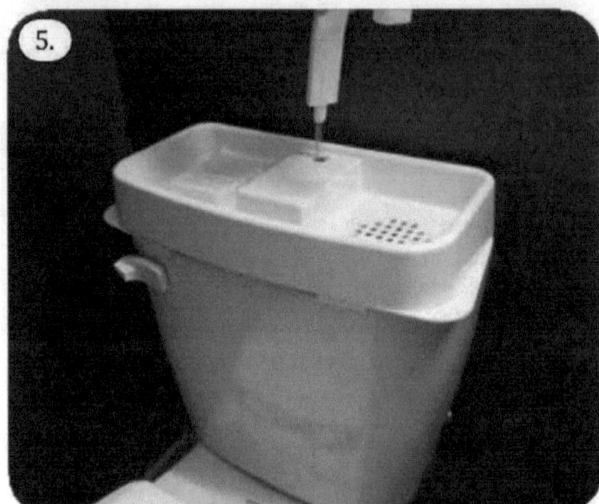

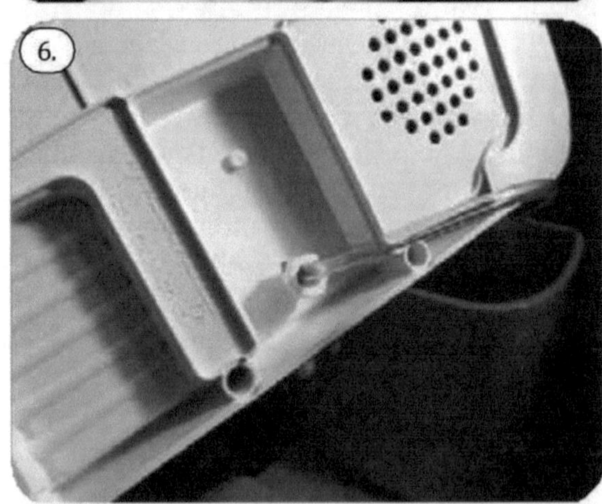

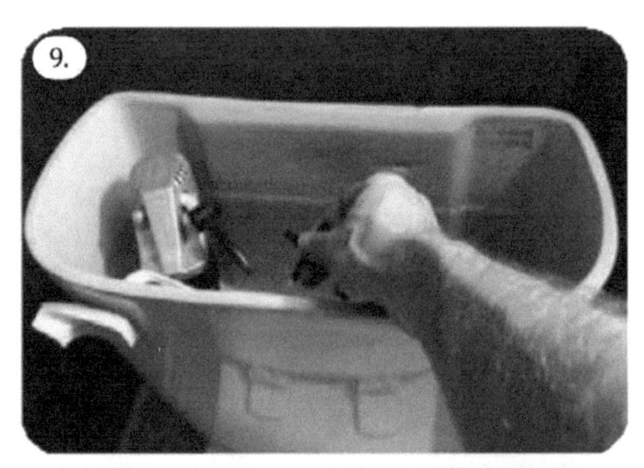

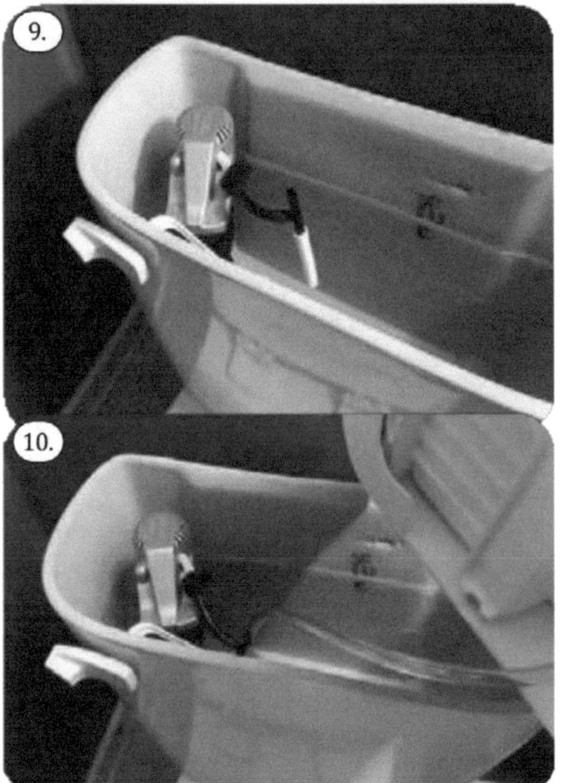

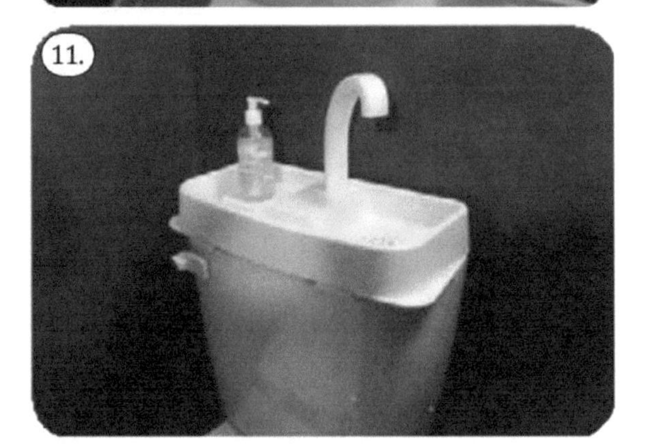

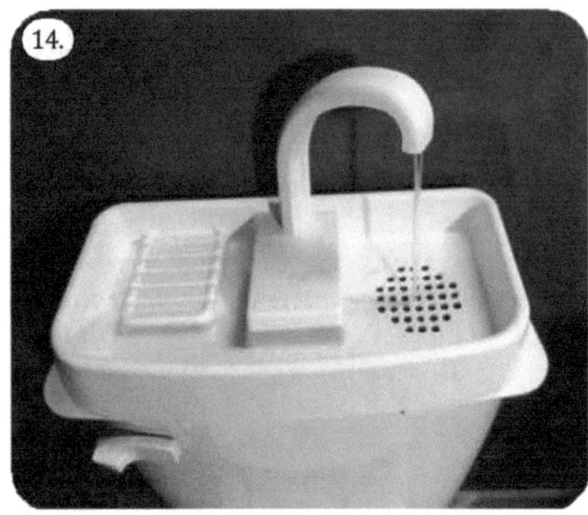

The more effectively you can manage your water resources the better life you will have off grid. This build is incredibly easy and practical, too. While the design might seem strange or awkward at first, it won't be long before you use this faucet as your defacto handwashing after using the bath- room.
The water you will save just from handwashing will make a massive difference over time. Not to mention the benefits to the planet of not wasting water.

How To Build A Rainwater Catchment System

MATERIAL COST  411.56 COMPLEX  DIFFICULTY 3 HOURS 

Fresh water falls from the heavens every so often, but normally this water flows through our gutters and into the ground. In an off-grid or grid-down scenario, we can not afford to let this resource slip away; we can and should harvest this rainwater to help us. Harvesting rainwater from our roof is as simple as catching it in barrels, but there is a way that we can use the space by our downspouts more efficiently.

What to Use for Rain Barrels

The best option for a rain barrel is a 55-gallon drum. While you can use steel drums, plastic drums are easier to handle and often easier to obtain. The drums designed for industrial use will be far stronger and made of higher quality plastic than the barrels often sold as rain barrels in big box stores. While you can purchase new drums, you may find some used plastic drums at a lower price, but you should try to lonely buy used food-grade drums.

Placement of the Rainwater Catchment System

Unless you want to drill holes into your gutters, you'll be placing your rainwater catchment system in very close proximity to an existing downspout. Try to put the system at a point that collects the most water from your roof.
Ultimately, you will have to determine the best way to attach this system's plumbing to the downspout or the connection in the gutter itself.

Building the Frame

When full, this frame is going to support almost 1500 pounds of water. Do not cheap out on materials for a build like this purchase the best wood and fasteners that you can. Also, take your time and make sure that all the joints are solid and the frame is square and level.

Tools

- Drill and drill bits and drivers
- Level
- Square
- Saw

Materials

- Three 55 Gallon Plastic Drums can be found at Home Depot for $90.00USD each.

www.homedepot.com/p/55-Gal-Blue-Industrial-Plastic-Drum-PTH0933/205845768

- Thirteen 2x4x8' boards are found at Home Depot for $6.55USD each.

www.homedepot.com/p/2-in-x-4-in-x-96-in-Prime-Whitewood-Stud-058449/312528776

- Three 1x4x8' boards which are $6.43USD at Home Depot.

www.homedepot.com/p/1-in-x-4-in-x-8-ft-Premium-Kiln-Dried-Square-Edge-Whitewood-Common-Board-914681/100023465

- #8 - 2 ½" Screws, which are $5.97USD a pound at Home Depot.

www.homedepot.com/p/Grip-Rite-8-2-1-2-in-Phillips-Bugle-Head-Coarse-Thread-Gold-Screws-1-lb-Pack-212GS1/100128995

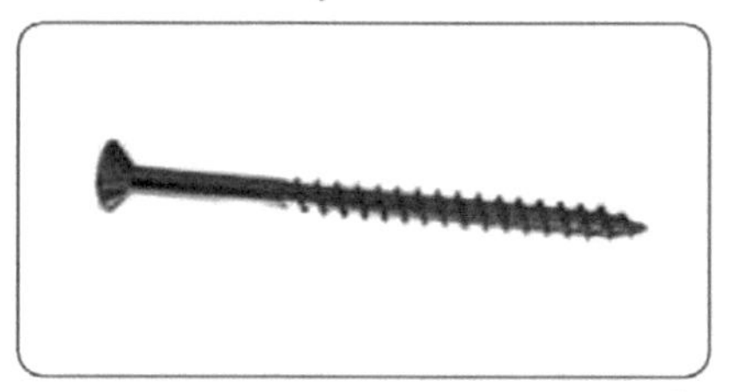

- #8 – 1 ½" Screws, which are $7.98USD for a pack of 100 at Home Depot.

www.homedepot.com/p/Everbilt-8-x-1-1-2-in-Zinc-Plated-Phillips-Flat-Head-Wood-Screw-100-Pack-801842/204275487

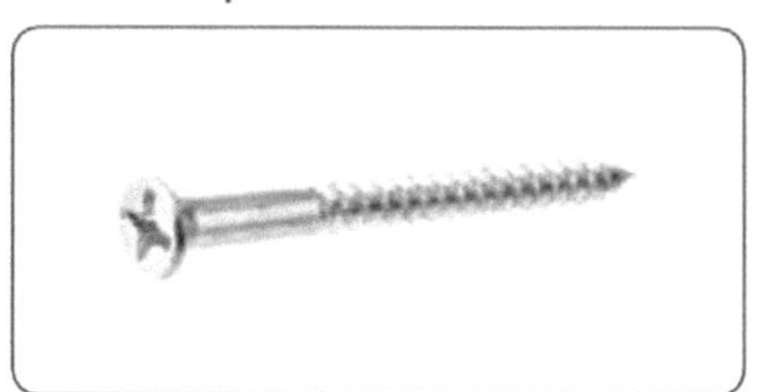

Lumber Cut List

2 x 4 x 8' Lumber

- 4 pieces @ 84 inches.
- 24 pieces @ 10 inch-
  - 12pieces@32inches.
  - 6 pieces @ 29 inches.

1 x 4 x 8' Lumber

- 6 pieces @ 35 inches.

Instructions

1. Cut all of your material as per the above cut list. If you need to modify the design to fit your space, sketch it out before cutting to deter- mine your true dimensions.
2. Layout and screw together the frame for the shelf on which the barrels will rest.
3. Attach two inside braces about 13 inches in from the outside edges.
4. Build the other two shelves on top of this one to make things easier.
5. Take one of the shelves and place a barrel on top of it, centring it. Place two of the angled 2x4s with offcuts of 1x4s up against the barrel as shown above.
6. Determine how far the 2x4s should be off the centre line of the shelf. This is the dimension that you will use to place all eight 2x4's on each shelf.
7. You will need to drill some holes to accommodate the screws about halfway through the boards.
8. Secure these 2x4's to the shelf, placing them at the dimension you determined in step 6.
9. Secure the 1x4s, drilling pilot holes to avoid splitting. Align the 1x4s as shown above and secure with 1 ½" screws to each angled 2x4.
10. Confirm that the barrel sits appropriately in the cradle.
11. Repeat for the next two shelves.
12. Take the bottom shelf to the location where it will be installed.
13. Prop the shelf up to the height that you want and level it.
14. Stack the remaining shelves on top.
15. Attach the 84" legs to the bottom shelf.
16. Lay the rack on its side to make spacing and securing the remaining shelves to the legs easier. In this case, I used an inside spacing of 24 ½".
17. Stand the rack up and slide in the drums.

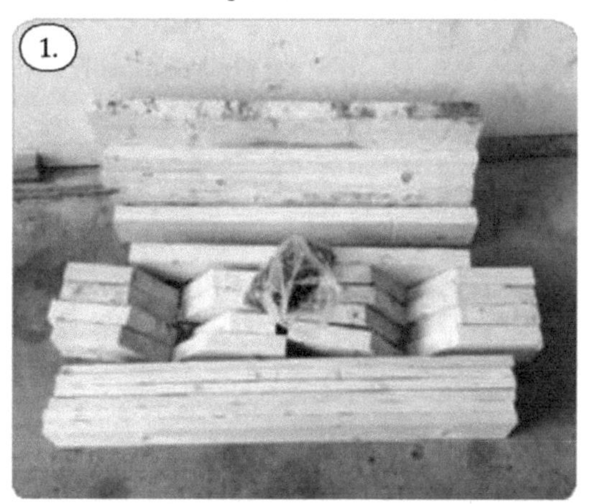

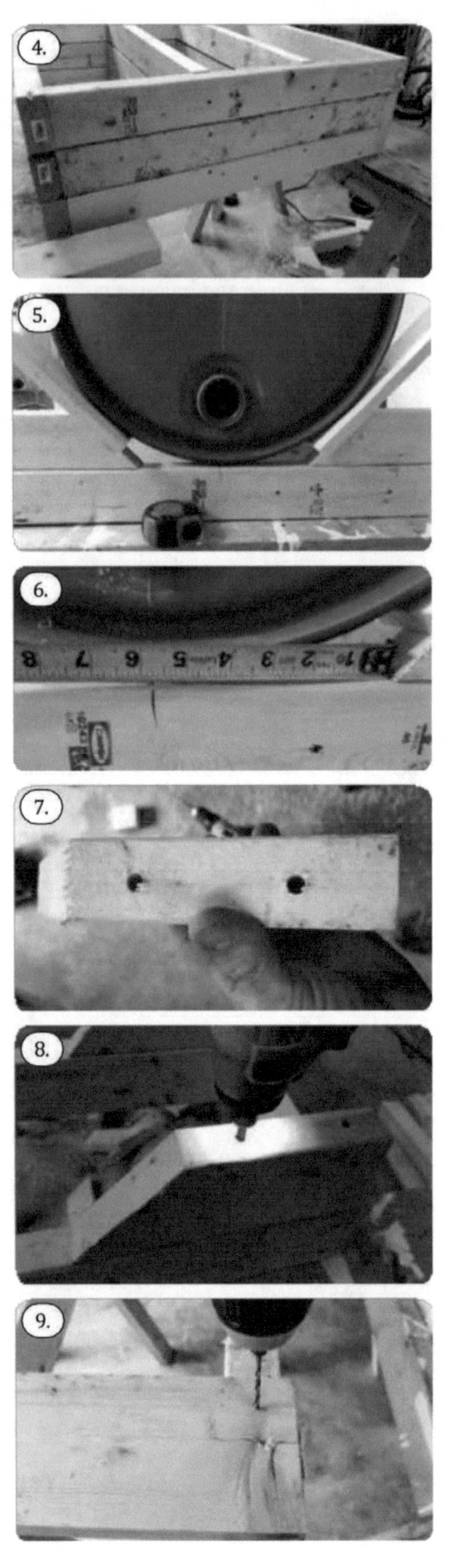
4.
5.
6.
7.
8.
9.

9.
10.
12.
13.
14.
15. 16.

Plumbing the Barrels

With the rack built and the barrels installed, it is time to plumb the system. There are a few different ways that you can go about moving water between the barrels. In this case, I used the ¾" NPT threaded knockouts in the barrel bung caps. If you can find larger fittings feel free to do so.

Tools

- Bung Wrench
- Saw
- Wrenches

Materials

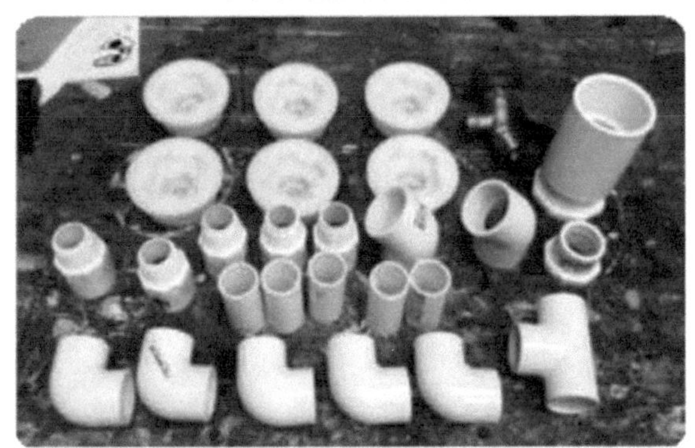

- Two feet of 1" Sch 40 PVC pipe which is $2.34USD at Home Depot.

www.homedepot.com/p/VPC-1-in-x-24-in-PVC-Sch-40-Pipe-2201/202300506

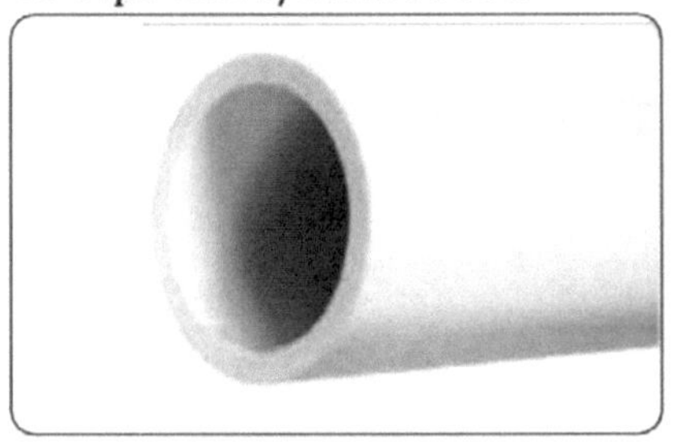

- Ten feet of ¾" Sch 40 PVC pipe which is $3.56USD at Home Depot.

www.homedepot.com/p/Charlotte-Pipe-3-4-in-x-10-ft-PVC-Schedule-40-Plain-End-DWV-Pipe-PVC-04007-0600/100348472

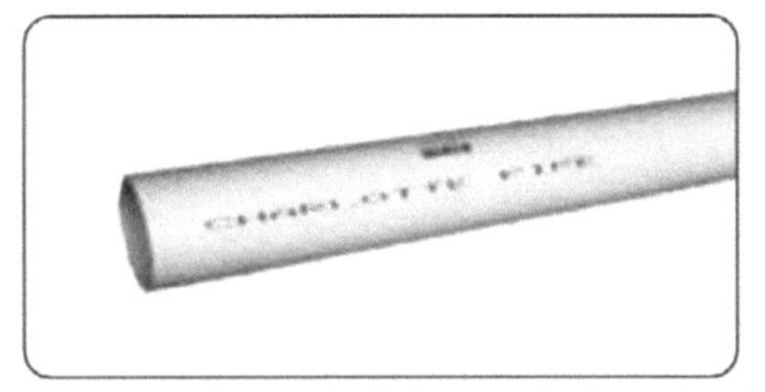

- One threaded hose bib, which Home Depot sells for $7.73USD.

www.homedepot.com/p/Everbilt-3-4-in-Heavy-Duty-Brass-MIP-x-MHT-Hose-Bibb-VHBCON-F4EB/312029306

- Five ¾" NPT to ¾" PVC Slip-on adapters can be found at Home Depot for $0.52USD each.

www.homedepot.com/p/DURA-3-in-x-4-in-Schedule-40-PVC-Reducing-Male-Adapter-MPTxS-436-341/203225030

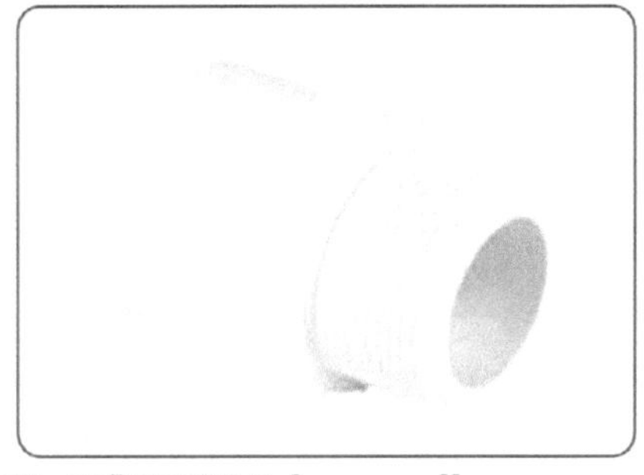

- Six ¾" PVC 90 degree elbows are sold at Home Depot for $0.52USD each.

www.homedepot.com/p/Charlotte-Pipe-3-4-in-PVC-Schedule-40-90-S-x-S-Elbow-Fitting-PVC023000800HD/203812123

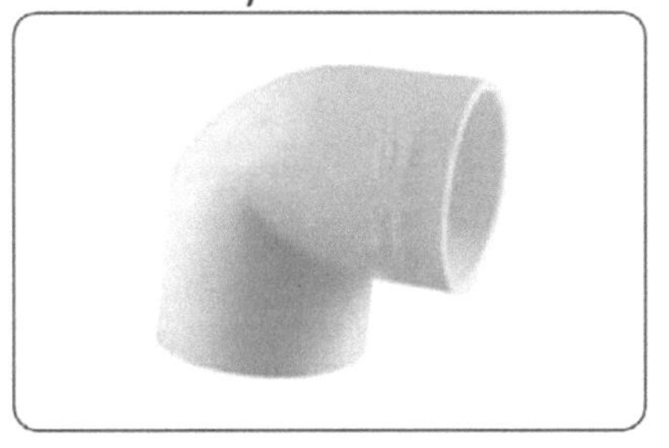

- One ¾" PVC slip-on tee can be found at Home Depot for $0.89USD each .

www .homedepot .com/p/C har lott e-Pipe-3-4-in-Schedul e-40-S-x-S-x-S -Tee-PVC024000800HD/203812197

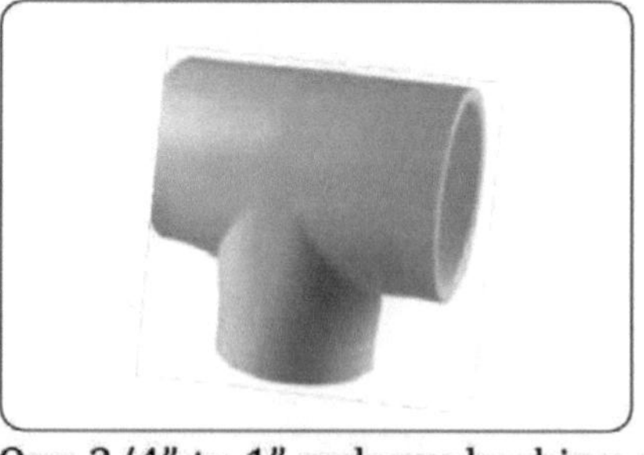

- One 3/4" to 1" reducer bushing, which Home Depot sells for $0.89USD.

www.homedepot.com/p/Charlotte-Pipe-1-in-x-3-4-in-PVC-Sch-40-SPG-x-S-Reducer-Bushing-PVC021070800HD/203811449

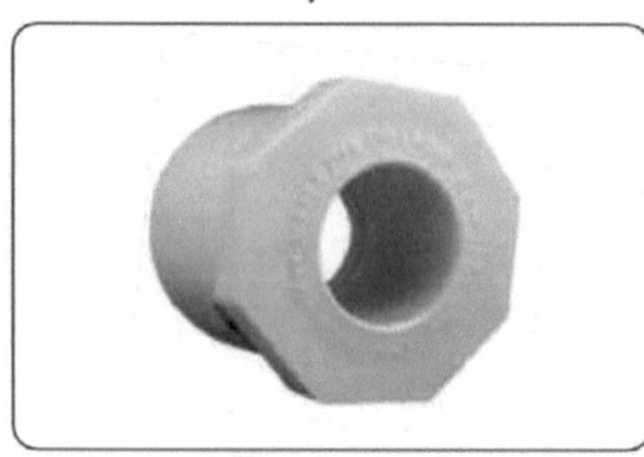

- Two 1" PVC 45-degree slip-on elbows are sold at Home Depot for $1.02USD each.

www.homedepot.com/p/DURA-1-in-Schedule-40-PVC-45-Degree-Elbow-C417-010/100345015

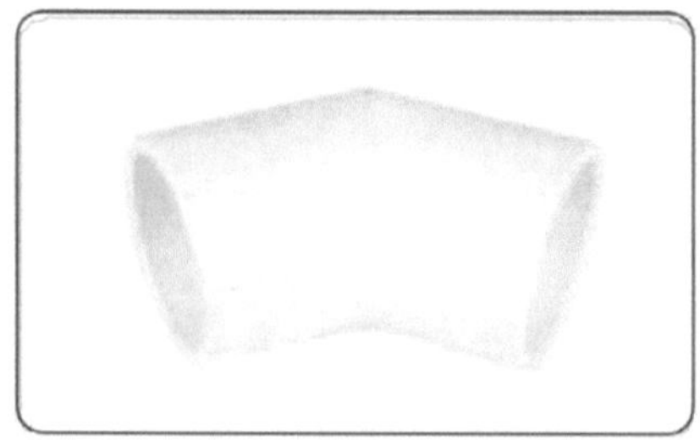

- You will need some miscellaneous fittings to adapt your downspout to the intake of the rain barrel system.

Instructions

1. Cut out the knockouts on each of the bung caps.
2. Install the bung caps into the drums.
3. To the bottom bung cap threaded hole attach the spigot.
4. To the rest of the bung caps, attach the ¾" PVC adapters and a 2-inch section of ¾" PVC pipe.
5. Install the 90-degree elbows and measure to determine the cut size for the ¾" pipe that will join the drums together.
6. Cut and install the pipes between the bottom and middle drum and the center and the top drum.
7. From the top bung cap, run a small length of PVC up above the drum's top and connect the tee joint. Run another short line off this tee joint perpendicular and connect a 90-degree elbow.
8. Run a line from this elbow down to the ground. This is your overflow line.
9. You will have to run pipes to either the downspouts or the gutter from the tee's open end. In my case, I used a 1" to ¾" reducer to widen the line to 1" PVC, then using lengths of 1" PVC pipe and two 45 degree elbows, I was able to make a run of pipe from the gutter to the drum.
10. I then used a 1 ½" to 1" reducer to widen this pipe system's inlet. You may have to use a different set of fittings to achieve the goal of moving water from your gutters to the rainwater catchment system.

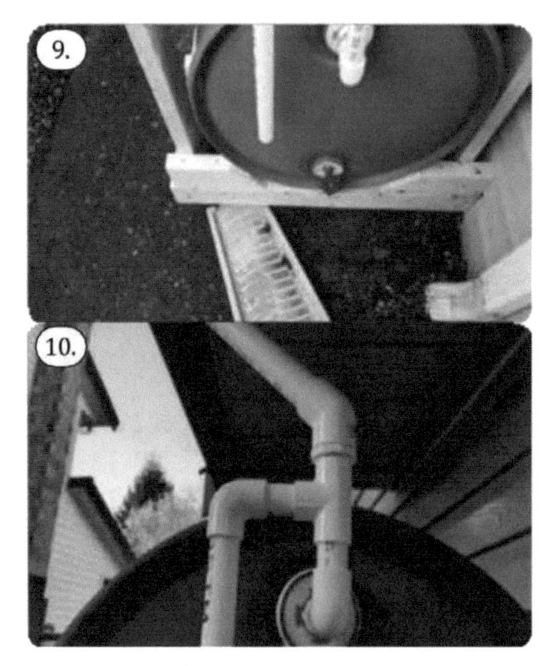

Variations

You may want to consider a few variations when designing your own three-barrel rain catchment system.

- If you can locate elbows with a hose barb instead of a slip-on fitting, you can use a clear plastic hose instead of PVC pipe, which will act as a sight glass and give you a visual of the water level in the drums.
- Instead of joining the drums with a PVC pipe section, you can have a continuous run of pipe from the lowest bung cap to the highest bung cap. Each of the bung caps connects to this run of pipe through PVC tees. Since all the drums and bung caps connect to one pipe run, the drums will vent air easier. This variation is good if you have a significant volume of wa- ter that runs through your gutters and downspouts.
- You should consider installing some mesh on the inlet to the rainwater catchment system to prevent dirt and debris from entering your barrels. The downside of this is that this mesh will become clogged and need to be checked and cleaned regularly.
- Rainwater has various uses around the garden and the house, both in a grid-down situation and in daily life. Three 55-Gallon drums stacked atop each other will give you access to 165 gallons of freshwater that you can use in your garden and around the home instead of using your drinking water stores. In a dire emergency, this water could be used as drinking water as long as it is filtered appropriately.

How To Build A Water Tank For Long Term Storage

EASY  DIFFICULTY

Water is the most critical element to survival besides air and shelter, so it must be one of our chief concerns when prepping for any emergency or disaster. The average human will not survive longer than three days without access to this precious resource, but our water requirement comes at a high price in terms of space and weight to store it.

How Much Water?

The average adult human needs about 1 gallon of water per day, and this is what we base our estimates on. Fortunately, the measurement of most water storage containers is Gallons so figuring out how many days a given container will supply us is very straightforward.

Start by forming a realistic view of how long you will need to live off your stored water. In my case, the most realistic scenario is a massive earthquake which would knock out services for at least a couple of weeks. My family has four people, so that would be four gallons of water per day. Right? Wrong. We can not forget about our pets, and my crazy chocolate lab also needs water to survive. To make matters easy, account for your pet's water needs the same as your own. In my case, I need to have five gallons of water on hand each day.

Five Gallons multiplied by 14 days equals 70 Gallons of clean and potable water. With each gallon weighing around 8.34 pounds, this means that two weeks of water for a family of four and a dog will equal about 584 pounds.

That is a lot of water that we need to find containers to hold it.

Container Options

There are a lot of options for vessels to store your emergency drinking water. Regardless of the type of container, the procedures for cleaning, sanitizing and storing remain very similar.

Food Grade 55 Gallon Plastic Drums

The best bang for your buck is a food-grade 55-gallon drum filled with water. You can store eleven days' worth of water for a family of five in a relatively small footprint. While these drums are great for water storage, they suffer from a couple of issues. If the drum ruptures or the water inside becomes contaminated, the result is a massive loss of clean drinking water. Also, a full 55-gallon drum of water is nearly 500 pounds, making it nearly impossible to transport effectively.

- You can purchase one at Amazon for $ 103.50 USD

www.amazon.com/Gallon-Plastic-Barrel-New-Factory-Fresh/dp/B07B6B2JP7/ref=mp_s_a_1_3?dchild=1&keywords=55+gallon+drum&qid=1616678249&sprefix=55+g&sr=8-3

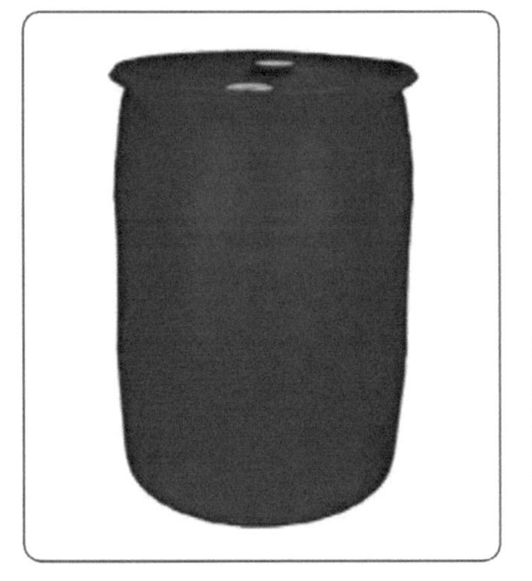

Jerry Cans

Water jerry cans come in various capacities, with the five-gallon size being common and a size that is not too heavy when full of water. Storing water in jerry cans is much the same as in the 55-gallon drum, but the size and shape of these cans mean that you can not keep the same water volume in the footprint of a 55-gallon drum. Jerry cans have the distinct advantage of being portable, and if you needed to bug out, they are much easier to take with you.

- Five-gallon jerry cans are available at Amazon for as low as $24.99USD I own a few of these, and I like that they are easy to manage and can be thrown into the back of the truck easily.

www.amazon.com/Scepter-04933-Water-Can-5-Gallon/dp/B000MTI0GA/ref=mp_s_a_1_3?dchild=1&keywords=5+gallon+water+jerry+can&qid=1616678393&sprefix=5+gallon+water+je&sr=8-3

- A 7 gallon Jerry can with a spigot is found on Amazon for $14.97USD. This one is nice because it is square and will not tip over. It can be laid over on its side, giving good access to the spigot.

www.amazon.com/Reliance-Products-Aqua-Tainer-Gallon-Container/dp/B001QC31G6/ref=mp_s_a_1_17?dchild=1&keywords=5+gallon+water+can+with+spout&qid=1616678715&sr=8-17

Soda Bottles

Soda or any other beverage bottle is not the ideal container to use, but if you are on a budget or have a surplus of them around, they can be an acceptable option. Soda bottles are good for people that are short on space, such as condo and apartment dwellers. Cleaning these bottles is difficult, and the water stored inside of them should be rotated regularly.

Disposable Water Bottles

You can get a flat of water at your local Costco for around six dollars, so it would seem like a good idea to have water stores of individual 500ml bottles of water. The problem with flats of disposable water bottles is that they were never designed to store water long term, and while the water inside of them will never expire, the plastic will break down over time.

That being said, rotating a flat or two along with your water stores is a good idea for use as barter items or to give to needy neighbours.

Designing a Water Storage System

It is important not to put all our eggs in one basket when it comes to long-term water storage. While 55-Gallon drums are an effective and efficient means of water storage, you need to have several different types and sizes of containers and store them in multiple locations.

Suppose you are going to base your water storage around 55-Gallon drums. In that case, it is imperative that you also have an assortment of smaller containers to allow for easier water movement from the drums to the locations inside your home that you require water. Using five-gallon jerry cans as your primary day-to-day water source and then refilling them from the 55 Gallon drums allows for easier access to water throughout the home. It also gives you the ability to take water with you during a bugout.

Think about how you will be accessing the water stores. If you store water in an outbuilding, you need to be very mindful of what scenarios might cut off your access to that water. In these situations, you'll need to take extra care to maintain a water supply where it is necessary to minimize the need for resupply.

Preparing Containers for Long Term Water Storage

When it comes to 55-Gallon plastic drums, you will need to purchase a few items along with them to clean, sanitize and fill effectively.

- Bung wrenchs are on Amazon for $16.98, and you will need this to remove and tighten the bung caps on a 55-Gallon drum.

www.amazon.com/Duda-Energy-dwrench-Aluminum-Standard/dp/B00950CICE/

- Regular unscented bleach with between 5.25% and 8.25% Chlorine you can find at your local grocery or big box store.
- Any dish soap will do for cleaning a new container.
- Food safe tubing is found on Amazon for $ 19.85 USD for 50 feet of tube.

www.amazon.com/50-ID-OD-Vinyl-Tubing/dp/B01LY3LSPE/

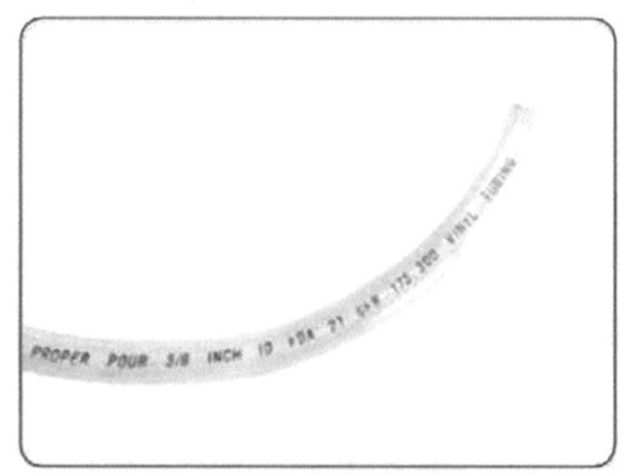

- Baking soda and vinegar are also available at your local grocery or big box store.

Even if you purchased brand new drums or jerry cans, you would still need to clean them inside and out. In the case of new containers, simply filling the container about a quarter of the way full with warm water and adding a generous squirt of dish soap is going to be more than adequate to clean it.

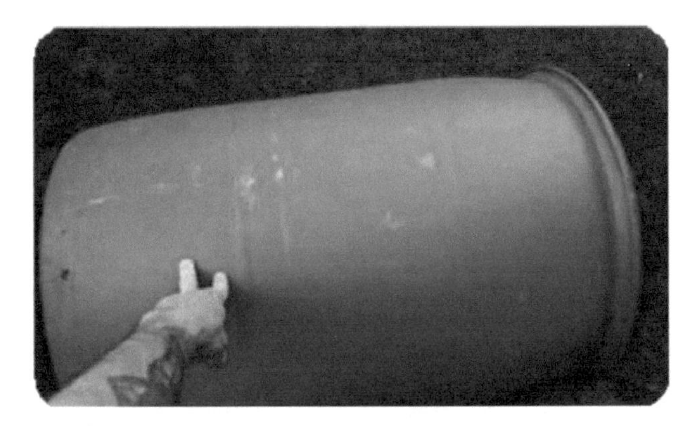

Seal up the container and roll it around, allowing the soapy water to cover the entire interior of the drum or jerry can.

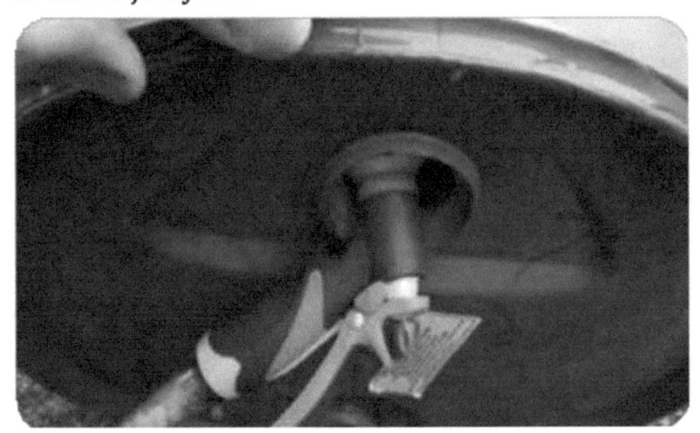

You will then need to rinse it out with clean water repeatedly until all the water you dump out of it runs clear and free of any soap or bubbles. This process will take a lot of time, do not rush it. Since we have not sanitized the container yet, it is ok to use a garden hose to rinse out the soap.

If there have been other substances stored inside the drum and dish soap is not doing a good enough job of cleaning the inside or there is an unpleasant ordor, you may need to use another cleaning method. Pour half a gallon of vinegar and a box of baking soda into a 55-gallon drum filled a quarter of the way with clean water. Allow this solution to work over every surface, rolling the drum around slowly, tipping upside down and letting it sit in various positions for long periods. You may want to let the solution work overnight.

Then rinse thoroughly and check for cleanliness. If this method does not work, the drum is probably not worth your effort, and you should consider using it as a rain barrel instead.

Sanitizing Containers

Once you have cleaned the container, you need to sanitize it before filling it. In the case of a 55-Gallon drum, add ¼ of a cup of unscented fresh bleach to the drum and then fill about a quarter of the way full of fresh, clean water.

Roll the drum around slowly, allowing the bleach to cover every surface of the container. This is a process that you need to take your time doing; it is better to be sure the entire inside is sanitary than sorry that eleven days' worth of water is spoiled.

After sanitizing, dump the bleach water out as best that you can. It is virtually impossible to remove every drop of water, but it is not a big deal if a little bleach water is left over.

Your drum is now ready for placement and filling.

Vertical Placement of 55-gallon Drums

The easiest way to store these drums is vertically, but this poses one significant issue. To get the water out of the drum, you will need to either siphon or pump the water out. When storing water vertically, consider the possibility that your pump will break and that you'll be forced to siphon.

No matter what, no water container should sit on the ground. There should always be a separation between the ground and the bottom of the container.

Horizontal Placement of 55-gallon Drums

To lay a drum horizontally, you allow for easier access to the water if you install a spigot. Since 55-Gallon drums have two ports and two bung caps, you can orient the drum so that one is at the bottom and the other is at the top. However, you also lose some water storage capacity since the fill port at the top will leave some headroom filled with air instead of water.
The best way to fill the drum is to install a fitting into the top bung cap with a short section of PVC pipe to fill the drum until it is as full as it will allow. Orienting the drums horizontally means that you need to build a stand for the drum that will elevate the drum high enough so that it can be used to fill any container that you need to fill and be strong enough to support 500 pounds of weight.

Filling a 55-gallon Drum

Once you have placed the drum in the location that you will store it, you'll have to fill them with clean water. Under no circumstances should you use a garden hose to fill any water container. Always use food-grade tubing to fill a container that will hold drinking water. If you do not have access to food-grade tubing or it is not practical, use your five-gallon jerry cans to ferry water to the 55-gallon drums.
To treat the water, you can use the same bleach that you used to sanitize the drum. If your local water supply is chlorinated, you do not need to treat your water. However, if you are unsure as to the condition of the city water, simply follow these guidelines:

- ¼ tsp. bleach per gallon for cloudy water.
- ⅛ tsp. bleach per gallon for water that is clear.

Once your drum, jerry cans, or bottles are filled, and treated seal them up and keep them away from extreme temperatures or sunlight.

Using Your Water Stores

When the time comes to use your water stores, it is a good idea to run the water through a filter in case something went wrong during the sanitization process leading to microorganisms running amuck through your water supply.
Even though water has no expiration date, you should habitually drain sanitizing and refill your water storage containers every six months. I use the interval for changing batteries in my smoke detectors for changing out my water. Every January and June, I empty, sanitize and refill all of my water stores which is a big job but will pay off when I need it.
Water storage is the least exciting prep but also is the most critical. With a little planning, a few containers, drums, and minimal supplies, you can secure your family's water needs through whatever disasters are on the horizon.

How To Make Drinkable Water Out Of Air

MATERIAL COST  129.27 EASY  DIFFICULTY 50 MINUTES 

Water is such a critical resource that human beings are hard-pressed to survive more than three days without it. Storing water is vital to our long-term survival and we should all have significant water stores. However, we also need to have other freshwater sources to round out our water preps. A creek, lake, or natural spring is ideal, but with a few easily obtained items, you can also extract fresh water from the air.

The humidity in the air around us is freshwater waiting for us to extract it. To do this, we are going to use an average household dehumidifier and a water filter. The dehumidifier fan pulls air into the unit and over coils that cool the air. The result is that the water in the air condenses on the coils forming water droplets. These drops of water are collected, and cool, dry air blows out of the dehumidifier. These drops of water collect in a tank that we empty when full.

Dehumidifiers are available in several sizes, from small tabletop units to larger ones requiring wheels to move around. When selecting a dehumidifier, our first consideration needs to be the power requirements to run them. Making drinkable water out of air in an SHTF or grid-down scenario only works if you have an excess supply of electricity to power them.

Consider using several smaller dehumidifiers rather than just a large one for this project. The reason for this is two-fold: a smaller unit draws less power, and with several of them in operation, you can decide which unit to power on and which to turn off depending on available power.

Another advantage to using several smaller units is that they often run on 12 volts of electricity. These units get electricity through a transformer plug that steps down the 120 volts of AC power from the wall outlet to 12 volts of DC power. If you are running your home on 12 volts DC power, you could, with a little ingenuity, power these units straight from your battery bank.

A major benefit of this project is that the machine constantly runs and draws water from the air without any input or effort from yourself. The device will work for you twenty-four hours a day as long as it has power and the air is humid.

The largest issue with devices like this is that they require electricity to operate and in an SHTF or grid-down situation, electricity will be an invaluable resource. As I stated before, it is a good idea to consider using several smaller units instead of one large unit.

The volume of water that these units produce versus the power you need to provide them may not make sense for your home during an SHTF or grid-down scenario. However, if you find yourself with excess capacity in your power system, then this project may be worthwhile to have running and to top up your water stores.

The final drawback is that its operation is very dependant on the weather and the levels of humidity in the air. If you live in a moist environment, then these units will work quite well, but if you live in a desert environment, then the amount of produced water will be significantly less. Also, the changing of the seasons will affect the amount of water these units produce. In the dryer months, the levels of water production will be noticeably less.

Safety of Dehumidifier Water

Water from a dehumidifier is not ready for human consumption. Since dehumidifiers pull water out of the air, any mould, viruses, bacteria, or other contaminants can tag along with the water drops. While the water in the tank seems essentially the same as distilled water, since the water was not exposed to heat high enough to kill pathogens, we must filter the water before drinking.

Choosing a Filter

There are many filters available on the market, but the filter we need for this build needs to filter water without a pump or suction. We want gravity to pull the water down through the filter and into whichever container we will store it in.

Any filter that can adequately filter water rendering it safe to drink out in the bush will work well. If the filter you use does not filter viruses and this is a concern for you, you should boil the water after filtering as an added precaution

Building Your Project

The construction of this project is very simple and requires only a couple of easily obtained materials.

Tools

- Saw
- Cordless drill with bits and drivers
- Sharp knife

Materials

- Dehumidifier, which I bought off Amazon for $39.99USD

www.amazon.com/dp/B09CFTBBX1/ref=redir_mobile_desktop?_encoding=UTF8&%2AVersion%2A=1&%2Aentries%2A=0

- Water Filters, like these filters that I found on Amazon for $24.95USD

www.amazon.com/Sawyer-Products-SP128-Filtration-System/dp/B00FA2RLX2/?th=1

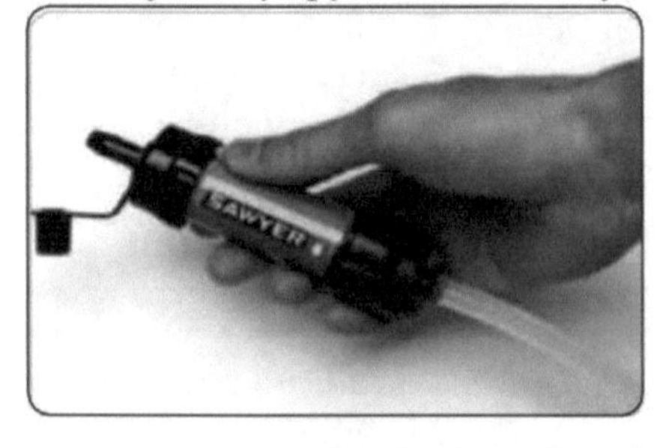

- Water Jerry, which I bought off Amazon for $ 31.65 USD and holds seven gallons

www.amazon.com/Reliance-Products-Jumbo-Tainer-Gallon-Container/dp/B000GKDFH4

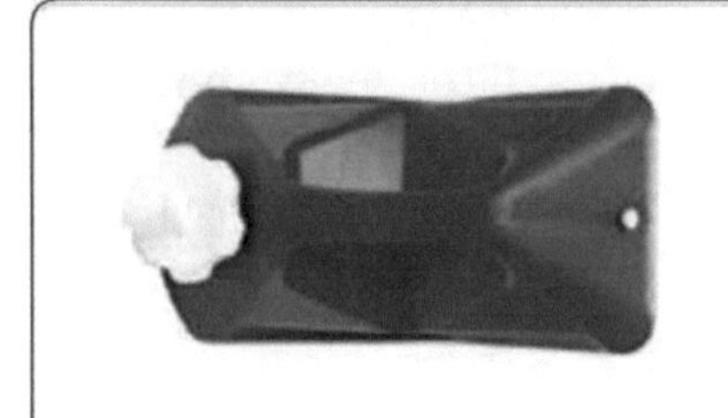

- ¾" PVC Pipe is available at Home Depot for $ 1.60 USD for a 24-inch length

www.homedepot.com/p/VPC-3-4-in-x-24-in-PVC-Sch-40-Pipe-22075/202300505

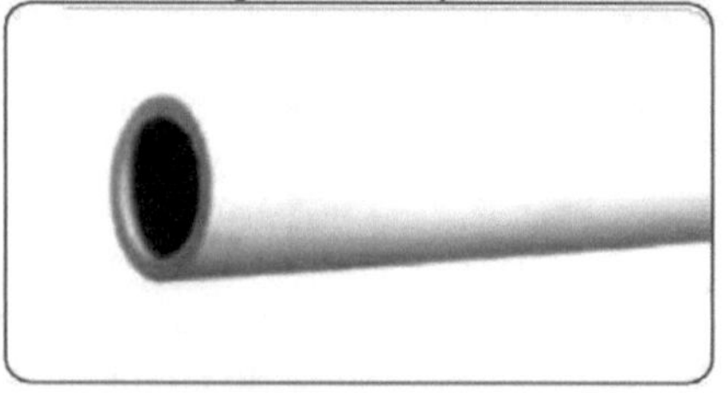

- ¾" PVC 90 degree Elbows are found at Home Depot for $ 0.59 USD each

www.homedepot.com/p/Charlotte-Pipe-3-4-in-PVC-Schedule-40-90-S-x-S-Elbow-Fitting-PVC023000800HD/203812123

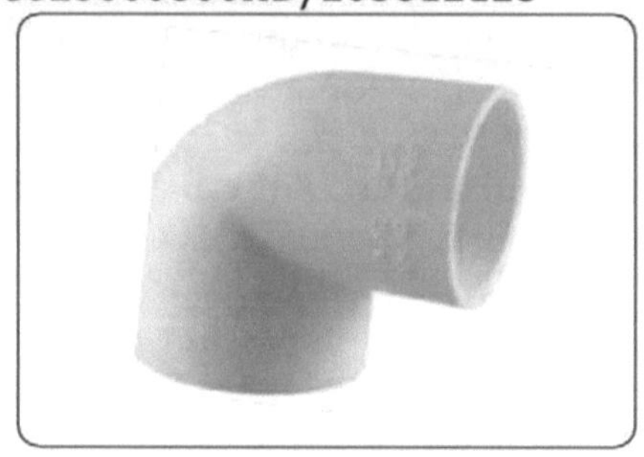

- #8 – 1 inch screws are found at Home Depot for $ 5.28 USD per pack of 100

www.homedepot.com/p/Everbilt-8-x-1-in-Zinc-Plated-Phillips-Flat-Head-Wood-Screw-100-Pack-801822/204275495

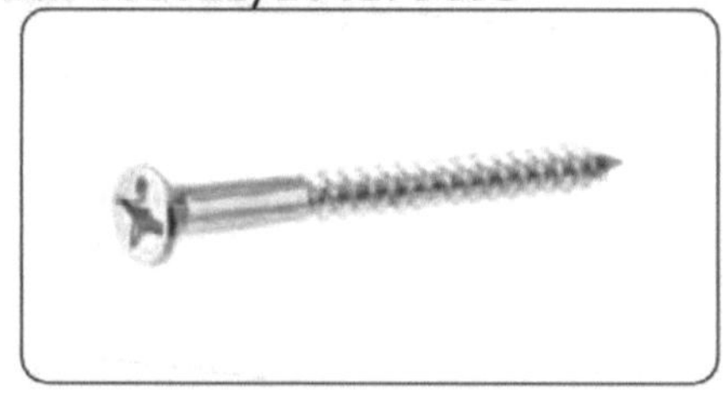

- Metal strapping can be found at Home Depot for $ 16.58 USD for a 100-foot roll

www.homedepot.com/p/Master-Flow-Perforated-Metal-Hanger-Straps-3-4HS/100396917

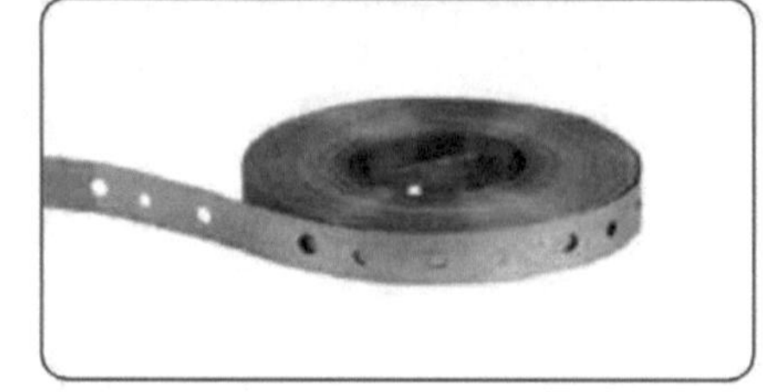

- 1x6 lumber is available at Home Depot for $ 8.04 USD for a six-foot boar

www.homedepot.com/p/1-in-x-6-in-x-6-ft-Premium-Kiln-Dried-Square-Edge-Whitewood-Common-Board-1X6-6FT/315221928

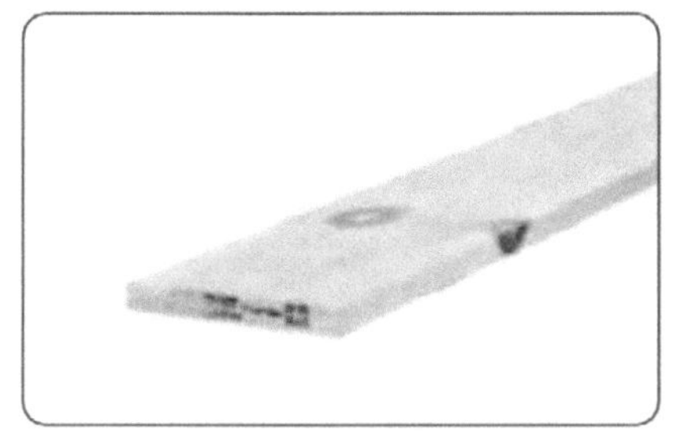

Instructions

1. Remove the tank from the dehumidifier and locate the port where the water flows into the tank. Most of these units have a switch that turns the unit off when the tank is full. This switch may need to be deactivated or avoided altogether to allow your build to work properly.
2. Cut a piece of ¾" PVC that is long enough to get water from the dehumidifier to a position above the jerry can or tank that the water will end up.
3. Attach a 90-degree elbow to each end of the ¾" PVC pipe.
4. Secure a length of ¾" PVC long enough to reach the jerry can or water tank below.
5. Align the open elbow with the water discharge port on the dehumidifier. You will proba- bly have to fashion a device to hold this pipe straight and in position so that the water flows out of the dehumidifier and through the PVC piping. In my case, I used some metal strapping and a scrap piece of 1x6 lumber to secure and lift the pipe, aligning it with the port.
6. I had to use a couple of small screws as levelling feet so that my pipe holder did not rock when in the dehumidifier.
7. For my filter assembly, I cut the bottom off a one-litre water bottle.
8. I then screwed it onto my Sawyer mini filter, creating the filter assembly.
9. I then placed the filter assembly into the mouth of my water jerry can and activated the machine.

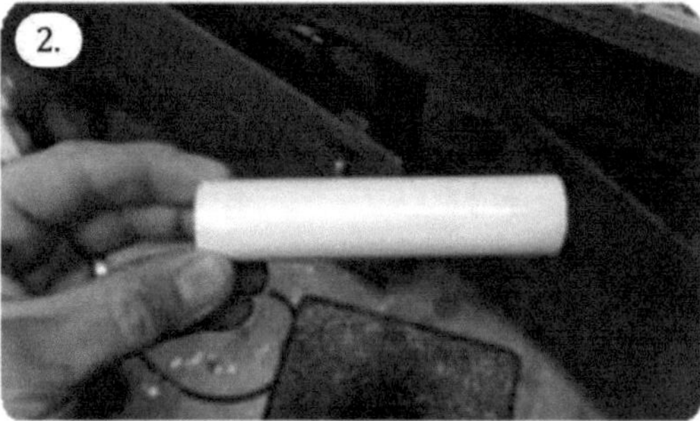

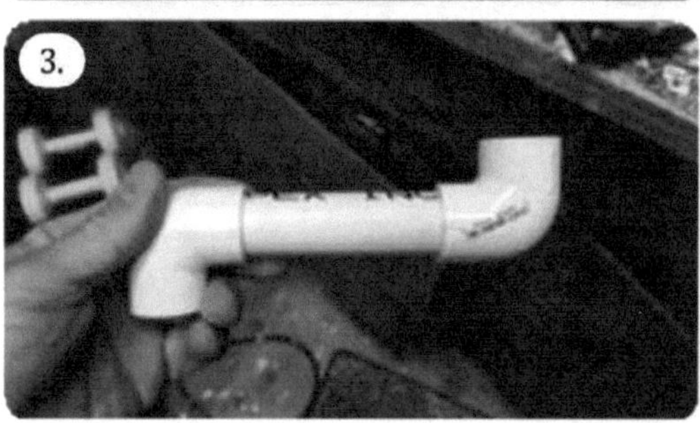

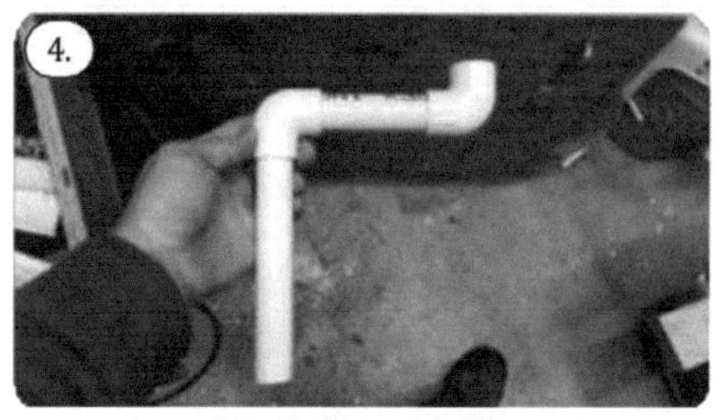

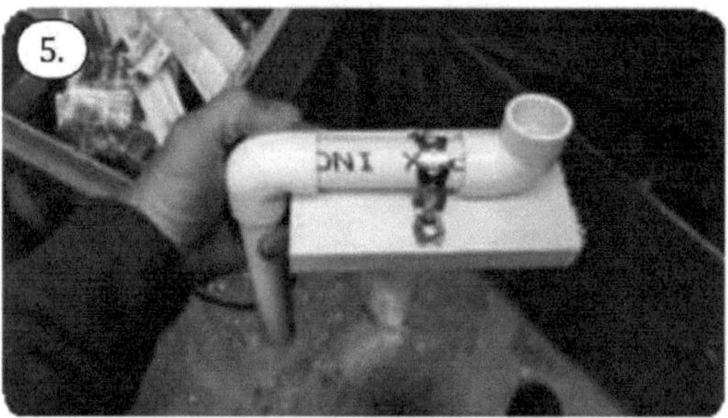

My Test Run

I set one of these units up in my garage and let it run for over 24 hours, producing about a quarter of a litre of clean filtered water. The problem is that the weather was fairly dry, so there was not a lot of humidity to work with. Had I placed the de- vice indoors in a room that sees higher humidity, such as a bathroom, I would expect the device to work more effectively.

I taste-tested the water and found it to have virtually no taste whatsoever.

This project is an effective way of pulling water from the air we breathe but suffers the signifi- cant drawbacks of requiring electricity to operate and depending on weather conditions. In no way should these devices be counted on as your sole water source, but you can find a place for one supplementing your water stores.

How To Purify Or Desalinate Water

MATERIAL COST $ 110.00 MEDIUM DIFFICULTY 3 HOURS

Water is so essential for life that the average human will perish without fresh drinking water in as few as three days. But unfortunately, another problem lurks within the precious liquid in areas where water is plentiful. Even crystal-clear water harbours dangerous pathogens and contaminants that make us ill if we drink it without purification. Many methods exist to remove pathogens and contaminants from water. Still, the problem with many of them is that they require filters or other equipment to render water safe to drink. Boiling is a fantastic option but requires a significant amount of fuel. Solar stills harness the sun's power and purify water through evaporation and condensation all by themselves. At the same time, we can carry on with other off-grid and survival tasks. Solar stills work by trapping evaporated water, allowing it to condense and be captured. When dirty water evaporates, only the water molecules are released, leaving behind all contaminants. The result is pure distilled water.

A solar still will remove all contaminants from water, leaving only pure distilled water to be collected, making a solar still a fantastic method of desalinating water.

Building a DIY Solar Still

Building a solar still is a very straightforward process, and the design can be modified to suit your needs. A few critical elements need to stay consistent, but the overall dimensions can be adjusted to suit your space and materials.

Tools Required

- Saw (Table saw rec- ommended)
- Drill with bits and drivers
- Tape measure
- Square
- Sharp knife

Materials

- Two sheets of ½" x 24" x 48" Plywood, which I found at Home Depot for $24.18 USD
www.homedepot.com/p/Sanded-Plywood-Com- mon-1-2-in-x-2-ft-x-4-ft-Actual-0-451-in-x-23-75- in-x-47-75-in-300896/202093833

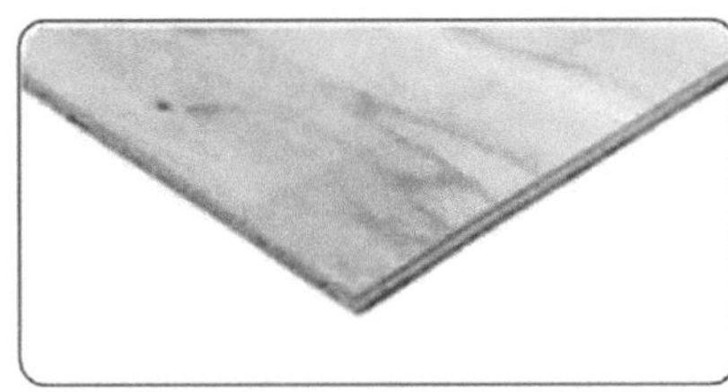

- Sheet of glass or clear plastic (I used an 18" x 24" Clear Acrylic Sheet, which Home Depot sells for $14.98 USD)
www.homedepot.com/p/OPTIX-18-in-x-24-in-x-0-093-in-Clear-Acrylic-Sheet-Glass-Replacement-MC-05/202038047

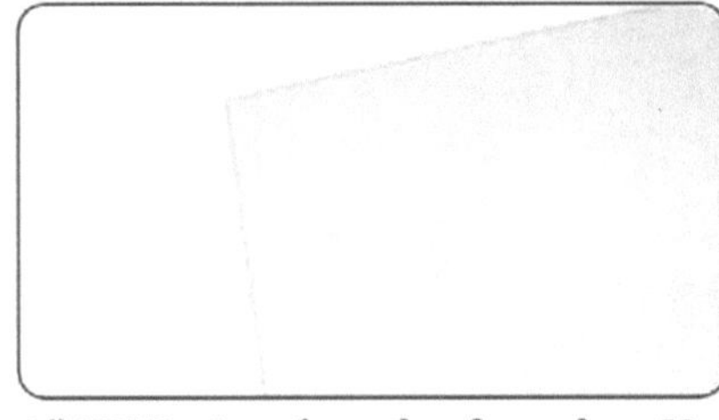

- 1" PVC pipe (can be found at Home Depot for $5.23 USD for a 24" length)
www.homedepot.com/p/VPC-1-in-x-24-in-PVC-Sch-40-Pipe-2201/202300506

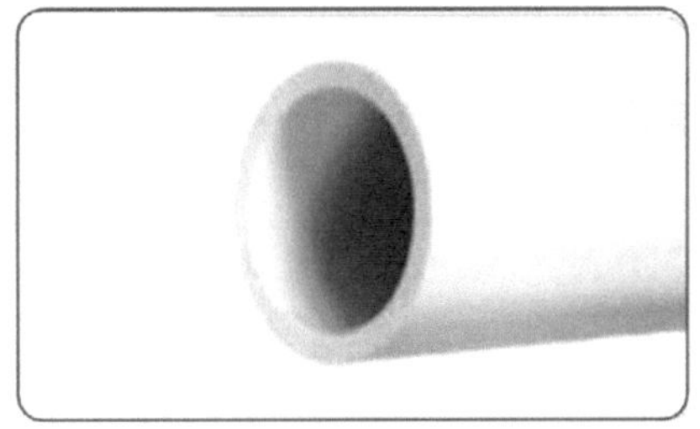

- 1" PVC 90-degree elbow (available at Home Depot for $1.42USD)
www .homedepot .com/p/C har lott e-Pipe-1-in-PVC-Sch-40-90-Degree-Elbow-PVC023001000HD/203812125

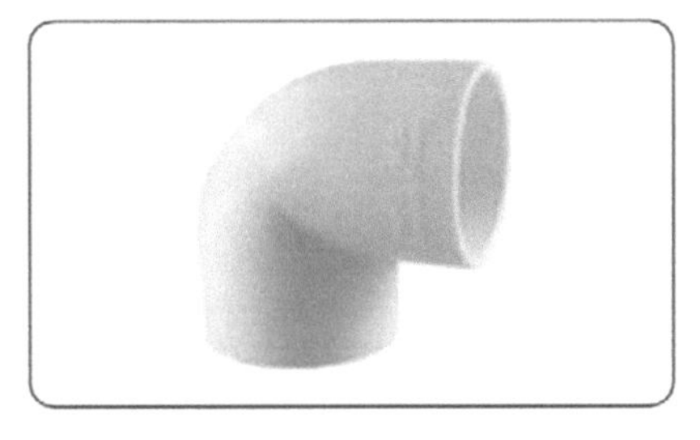

- #8 x 1" Screws (can be found at Home Depot for $6.30USD per pack of 100)

www.homedepot.com/p/Everbilt-8-x-1-in-Zinc-Plated-Phillips-Flat-Head-Wood-Screw-100-Pack-801822/204275495

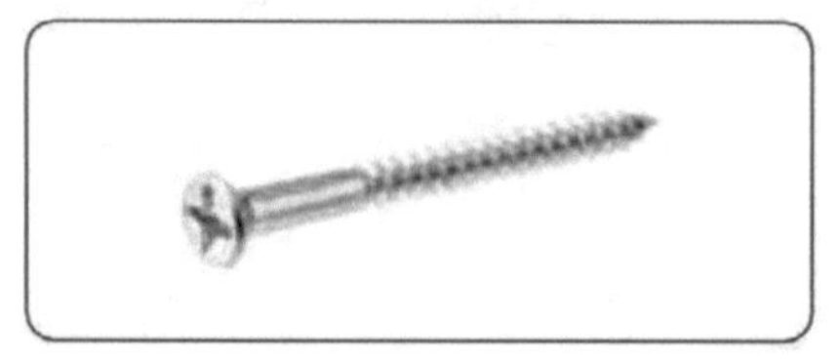

- I used a regular door hinge, which can be found at Home Depot for $3.82USD each

www.homedepot.com/p/Everbilt-3-1-2-in-x-1-4-in-Radius-Satin-Nickel-Door-Hinge-14985/202558077

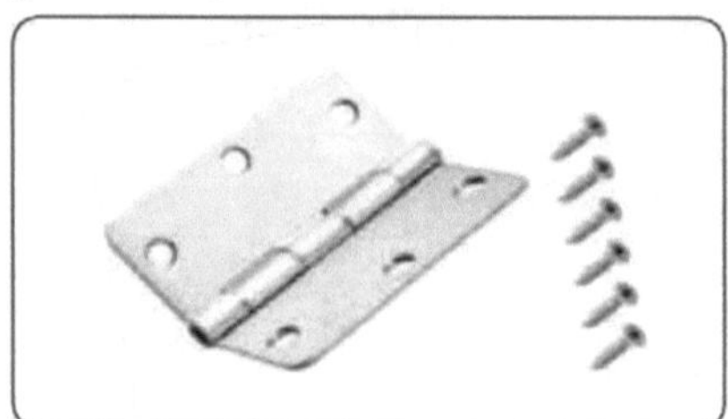

- Double Hinge Hasps (are sold at Home Depot for $4.18USD each)

www.homedepot.com/p/Everbilt-3-in-Zinc-Plated-Double-Hinge-Safety-Hasp-15128/202033920

- ½" Rigid foam insulation (available at Home Depot for $8.98USD for a 4' x 8' sheet)

www.homedepot.com/p/R-Tech-1-2-in-x-4-ft-x-8-ft-R-1-93-Insulating-Sheathing-320810/202533656

- Clear silicone (available at Home Depot for $11.98USD)

www.homedepot.com/p/GE-Advanced-Silicone-2-10-1-oz-Clear-Kitchen-and-Bath-Caulk-2708923/100663319

Instructions:

1. Determine the size of the solar still. There are a couple of ways to do this, one is to size the still based on the size of the trays that you will use to hold the dirty water, and the other is to size it based on the dimensions of the glass pane that you can source. In my case, I didn't want to cut the transparent sheet of plastic that I bought, so I sized the still based on its dimensions.
2. One of the still's essential aspects is that the glass is sloped down towards one end of the still. The angle of this slope is not critical and what I decided on was a 4" drop over the 24" length of the plastic sheet. I decided to make the back end of the still 12 inches tall, so I marked out 12 inches. I marked a point 8 inches up from the bottom and used the glass pane to mark where they would intersect.
3. I then cut my two side pieces using a table saw.
4. Next, I set up my table saw to cut a groove into the plywood about 1/8" to 3/16" deep.
5. Cut a groove into the plywood on one side of the diagonal edge of the plywood side pieces. This groove will hold the glass, so you may need to widen it depending on the thickness of the glass pane. Remember that these side pieces will be handed parts, and you need to consider which side is which when cutting.
6. Test fit the glass and measure for the bottom and the two end pieces.
7. Cut these parts.
8. Remove the glass and glue and screw the sides to the bottom. Secure the short end with glue and screws as well.
9. Temporarily secure the tall side with a screw or two.
10. Line the inside with foam insulation, keeping in mind that the tall end will be a hinged door.
11. Glue this insulation into place.
12. You can choose to paint the inside black with high-temperature paint, which may allow for greater heat absorption. Still, I can not say whether there would be any contamination of the water vapour from the paint. Therefore, I decided not to paint my solar still. If you choose to paint yours, allow it to thoroughly dry and cure according to the manufacturer's

directions.

13. Install a hinge and hasp on the tall piece and remove the temporary screws.
14. Drill a hole large enough to fit the 1" PVC pipe snugly. This hole should be positioned so that it is as close as practical to the glass top of the still.
15. Test fit the pipe and mark the point at which it enters the inside of the solar still.
16. Split the pipe in half only to the dimension equal to the inside dimension of the still. The goal is to have the section of pipe inside the still be open at the top.
17. Insert the pipe into the hole and position it so that there is enough slope that water will easily flow towards the outlet. Next, run a bead of silicone along the edges of the section of the pipe. Then slide the glass into place and run a bead of silicone around the perimeter of the glass.
18. Attach a 90-degree elbow to the outlet.

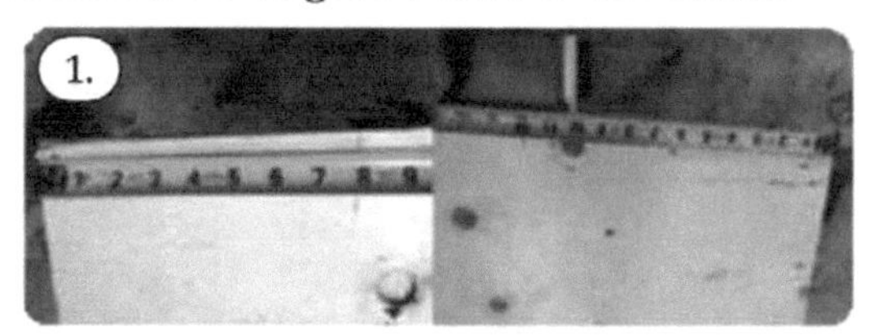

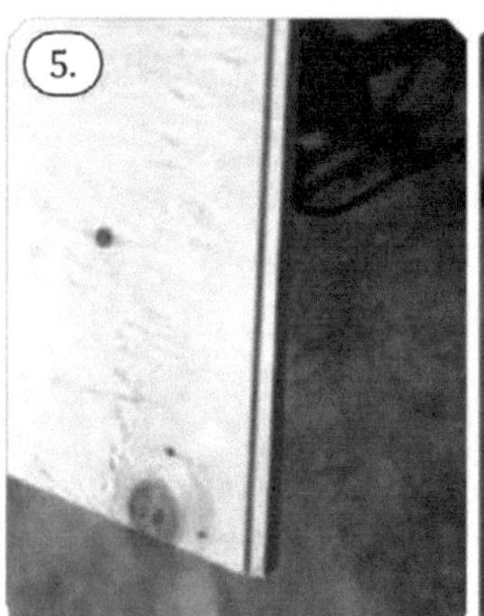

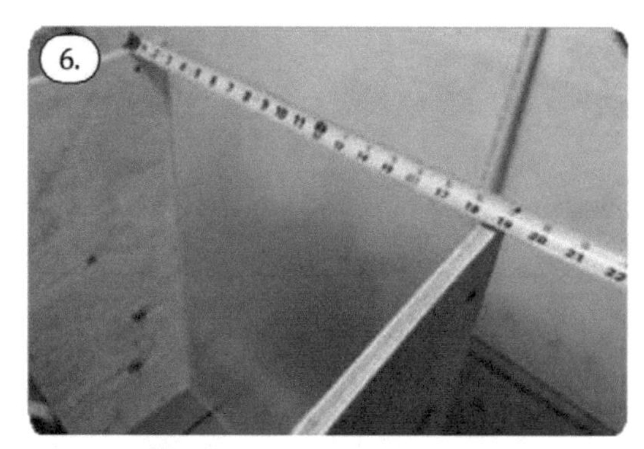

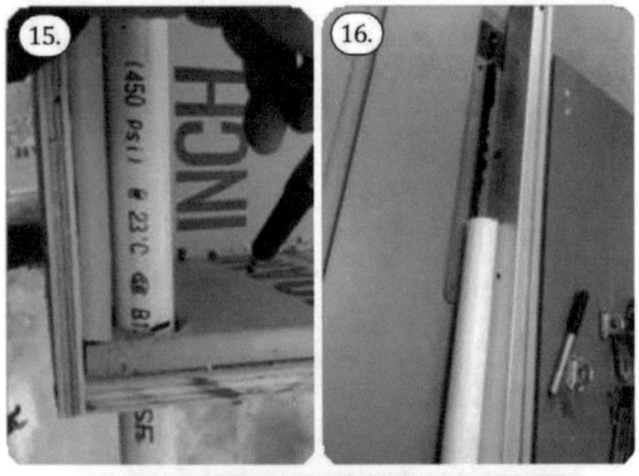

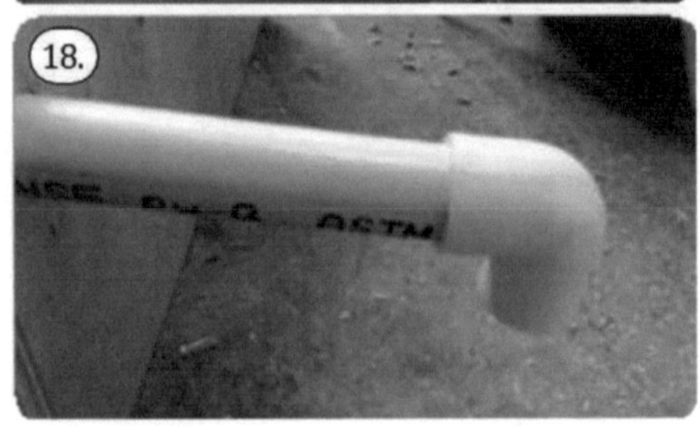

Using the Solar Still

To use this solar still, all you need to do is place it in direct sunlight and then load trays of dirty water into the body of the Still.

As the water evaporates, it will condense on the glass and trickle down into the trough. The clean water will then flow down into the container placed by the outlet to collect the clean and distilled water.

Considerations

When building a solar still, there are a few considerations that need to be made:

- While this solar still effectively removes impurities from water, it also removes all the good minerals from the water.
- A solar still requires the heat from the sun to work, so if you live in a climate or area with limited exposure to sunlight, this may not be a worthwhile investment in water security.
- A solar still will only produce small amounts of water. It would have to be scaled up significantly to be counted on as a reliable water source. However, it is an excellent way to desalinate water without requiring the burning of fuel to boil water.
- Care should be taken to ensure that the solar is still as airtight as possible. Any vapour that escapes is less than you can collect.

Water security is a paramount consideration in any preparedness plan. However, like all other preps, it is essential to have redundancies in place. Solar stills effectively remove everything aside from water molecules but require practical exposure to sunlight to work, limiting their deployment. Therefore, solar stills are best used as a supplementary water purification system or for desalinating water, rather than as your primary purification method.

How To Build A Hydroelectric Generator

MATERIAL COST  58.17 COMPLEX  DIFFICULTY 3 HOURS 

Hydroelectric power is in wide use throughout the world, for a good reason. It is a truly renewable resource since as long as the water is flowing, you have energy that you can turn into electrical power. Most designs for DIY hydroelectric power plants involve a DC motor or a vehicle alternator, but this build will demonstrate the mechanics of how a hydroelectric generator functions imparting some knowledge in the meantime.

How Is Hydroelectric Power Made?

Water moves through a turbine which spins, creating mechanical energy. This mechanical energy is converted to electrical energy by way of a generator. We are left with a few options for those of us who wish to DIY our hydroelectric generators. In some cases where there is a significant water flow, you can use a car alternator or DC motor attached to the turbine. The water will turn the turbine which spins the shaft of the alternator or DC motor, which will convert that mechanical energy into electrical energy.

In this build, we will construct a generator from scratch. While it may be easier to use a DC motor or alternator, preppers and survivalists need to understand the inner workings of devices like these. Only by understanding how they work, can we modify and repair our power sources when they inevitably require servicing.

How a DIY Hydroelectric Generator Generates Electricity

The generator we will be building uses coils of magnet wire and some strong rare earth magnets to turn the mechanical energy into electrical power. How this works is that the magnets will be fixed to a disc attached to the turbine's shaft (the rotor). This disc spins over a stationary disc (the stator) that holds the coils of wire. What happens is that the magnetic fields are picked up by the coils of wire which in turn generates an electrical current in those coils of wire. This is a dramatic oversimplification, but it is basically what is happening when the magnets spin over the coils.

Benefits and Issues with Hydroelectric Power

No method of generating electricity is without drawbacks or benefits, and hydroelectric power generation is no exception.

The main benefits of hydroelectric power are that it is clean and renewable. The process generates zero emissions and will generate power as long as the water is flowing. Hydroelectric power is also very quiet since the only sounds are the spinning of the turbine.

Some drawbacks are that when the water freezes, you lose your source of power. Along with freezing, the water source may dry up or have an insufficient flow to generate any usable power. Another issue is that your water source may be a significant distance from your home, in which case you will need to run power lines from the generator to your home. This means that there will be some loss of power along the way.

While a gas or diesel generator can produce several thousand watts of power and 120 or 220 volts of AC power, a homemade hydroelectric generator will not see high wattages like a gas generator. With some experimentation and ingenuity, you can use a generator like the one I describe to generate voltages around 12 volts.

How to Use a Hydroelectric Generator

Powering anything directly from the generator is never a practical solution. Instead, you'll need to set up a battery bank that uses the hydroelectric generator to keep the batteries charged. To do this, you will need to determine how large a battery bank you will need to power the devices and how much input power you will need to keep the batteries topped up.

Then you can build a hydroelectric generator, after which you need to test it and come up with a baseline level of power that you can expect it to generate. Remember that these generators are

running 24 hours a day, effectively charging your batteries while you sleep. If you find that the power generated by one of these generators is too low, then consider installing multiple units, all delivering power to your battery banks.
As for batteries, there are far too many options to cover here effectively, but you will need to look into charge controllers and transformers to convert AC power into DC power when setting up your off-grid power system.

Building a Basic DIY Hydroelectric Generator

This is a very bare-bones hydroelectric generator but will serve as a fantastic foundation to build on when considering hydroelectric power. While the generator shown here does not produce a significant amount of voltage, with a few additions and modifications, you can achieve voltages of more than 12 volts.

Tools

- Saw
- Soldering iron
- Square
- Tape measure
- Side cutters
- Razorblade or emery cloth
- Glue

Materials

- 22AWG magnet wire is sold on Amazon for $ 20.98USD per one-pound spool.

www.amazon.com/BNTECHGO-AWG-Magnet-Wire-Transformers/dp/B07DYMMYSK/

- Rare earth magnets are sold on Amazon for $16.99USD per package of eight.

www.amazon.com/Super-Strong-Neodymium-Magnets-Powerful/dp/B072KDBJWC/

- 3/16" Hardboard panels are sold at Home Depot for $17.29USD a sheet.

www.homedepot.com/p/Hardboard-Tempered-Panel-Common-3-16-in-x-4-ft-x-8-ft-Actual-0-175-in-x-48-in-x-96-in-832780/202404545

- ¾" PVC pipe is available at Home Depot for $ 1.60USD for a 24" length.

www.homedepot.com/p/VPC-3-4-in-x-24-in-PVC-Sch-40-Pipe-22075/202300505

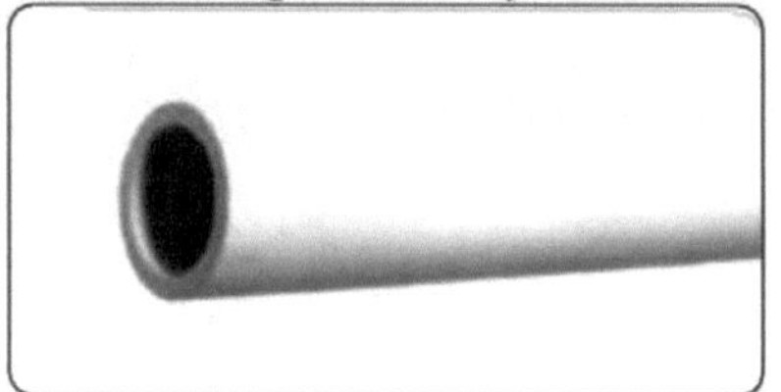

- ½" PVC pipe sold at Home Depot for $1.31USD for a 24" length.

www.homedepot.com/p/VPC-1-2-in-x-24-in-

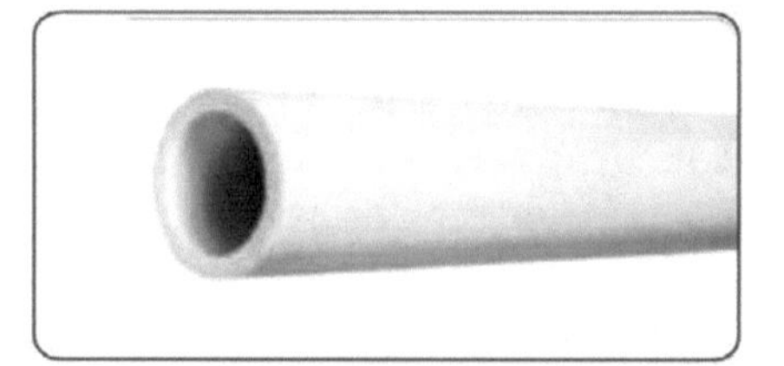

Instructions

BUILDING THE STATOR

1. Wind coils of 22AWG magnet wire around a form such as ¾" PVC pipe. Leave a tag end of a few inches to accommodate joining to another coil. Wrap the wire 200 times, leaving a tag end of a few inches. Scrape the enamel off the free ends of the wire so they can be soldered together later.
2. Cut two discs from the hardboard of a diameter that will accommodate the number of coils and magnets you have chosen to use.
3. Drill a hole in the disc centre that will hold the coils large enough to accommodate the ¾" PVC pipe. This disc is what is known as the stator.
4. To attach the coils to this disc, drill holes in the disc on either side of the coil and in the centre of the coil. These holes should be appro-

priately placed and sized to thread zap straps through.

5. Arrange the coils so that the electron flow will alternate between clockwise and counter-clockwise as it moves through the coils. This is a very important step. Trace the direction that the wires are coiled from the input wire to the output wire to determine whether the coil is clockwise or counterclockwise.
6. Solder the free wires of the coils together, joining input and output wires. Leave the last set of wires free.
7. Drill two holes into the stator and install a couple of machine bolts to attach the loose wires.
8. Connect wires to the opposite side of the stator to the machine screws.
9. Test the resistance of the coils through the machine screws. You should see a reading of fewer than 10 ohms.

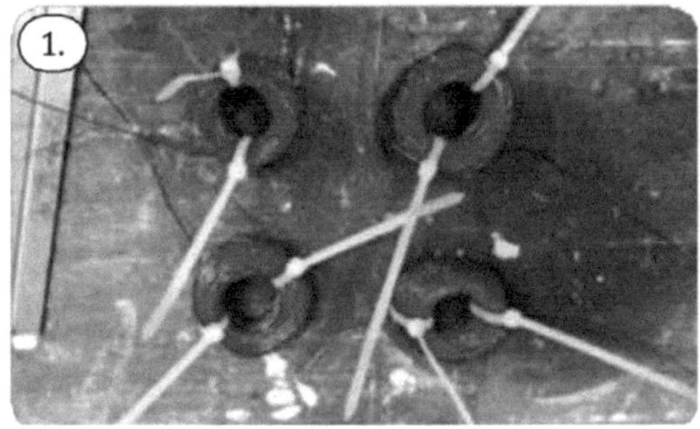

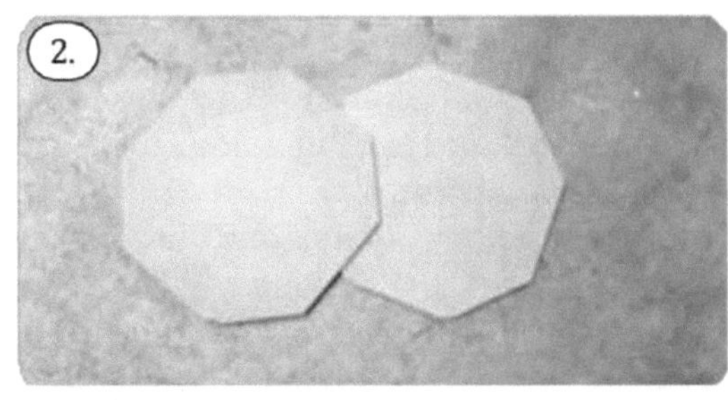

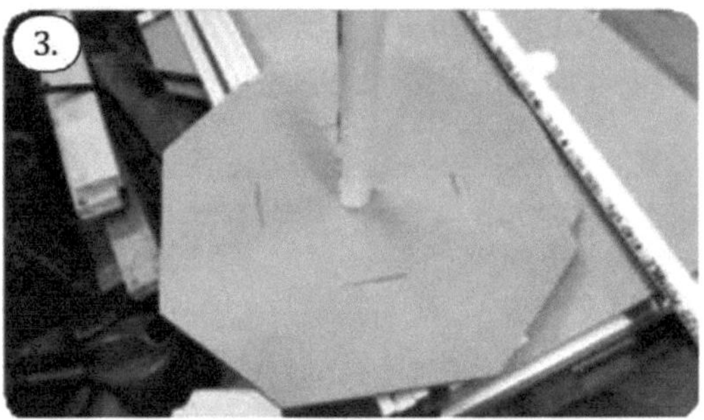

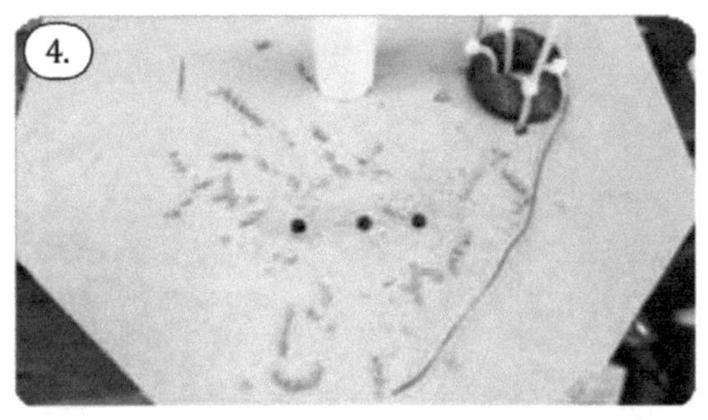

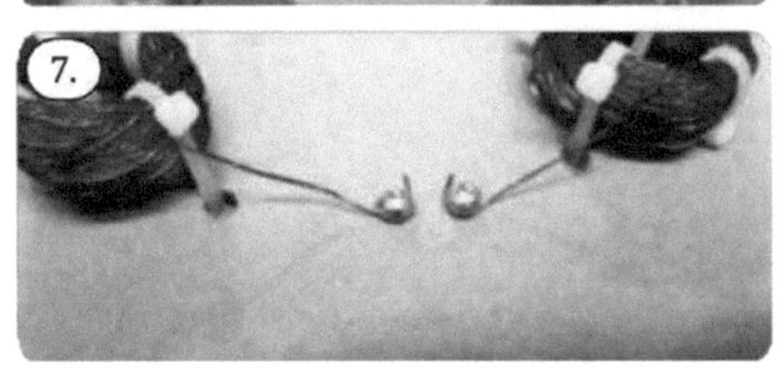

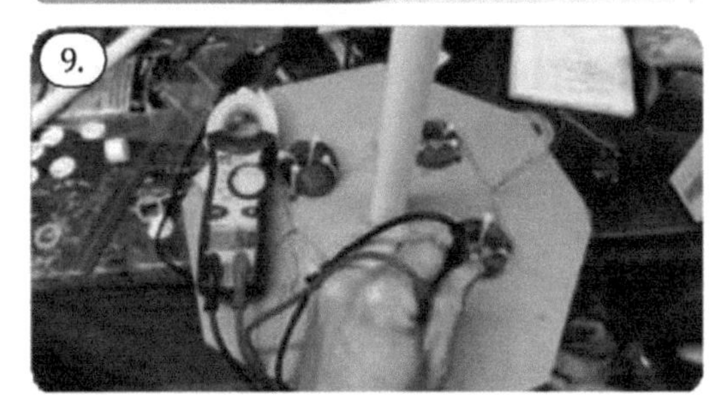

BUILDING THE ROTOR

1. Drill a hole in the centre of the disc that will hold the magnets sized to accommodate the ½" PVC pipe.
2. Glue or use double-sided tape to attach the magnets to this disc. Arrange them so that the North and South poles will alternate. Since my package of magnets contained eight magnets, I used all eight on my rotor.
3. Slide the disc onto the ½" PVC pipe and glue it in place. This pipe will be the axle of the generator, so make the length long enough to slide into the section of pipe attached to the stator and leave enough on the other side to accom-

modate the water wheel that we are about to construct

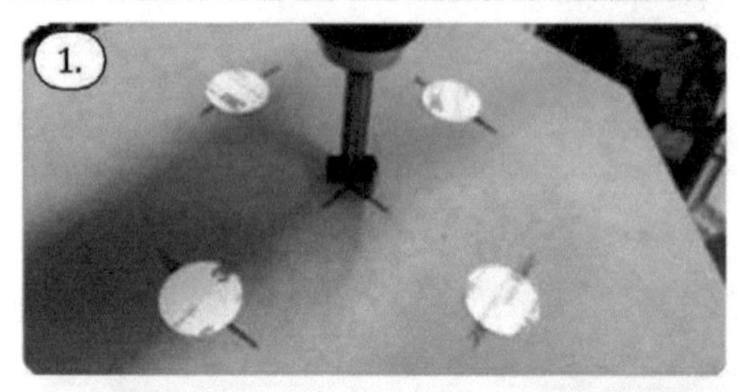

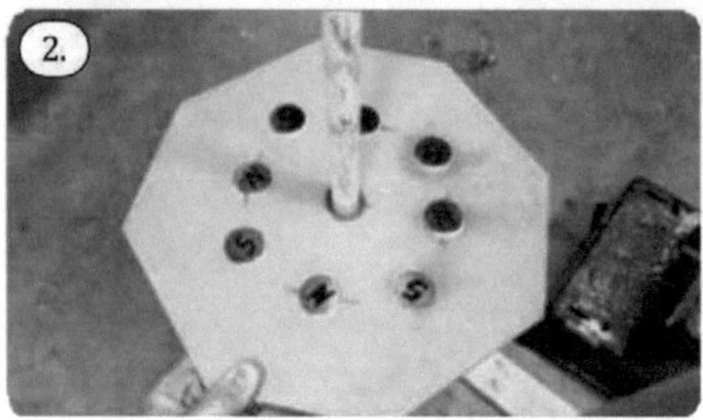

WATER WHEEL CONSTRUCTION

1. Use the remainder of the hardboard to cut at least six rectangles three inches by five inches and two discs around 7 inches in diameter. If you want to build a larger water wheel, adjust your dimensions accordingly.
2. Cut a piece of ¾" PVC to six and a half inches.
3. Drill a hole to accommodate the PVC pipe that you just cut in the centre of each disc for your waterwheel.
4. Slide the discs onto the PVC pipe.
5. Glue and secure the rectangle pieces to the PVC pipe and between the discs to form the paddles of the water wheel.
6. Slide the water wheel onto the axle and secure it to the axle using a screw.

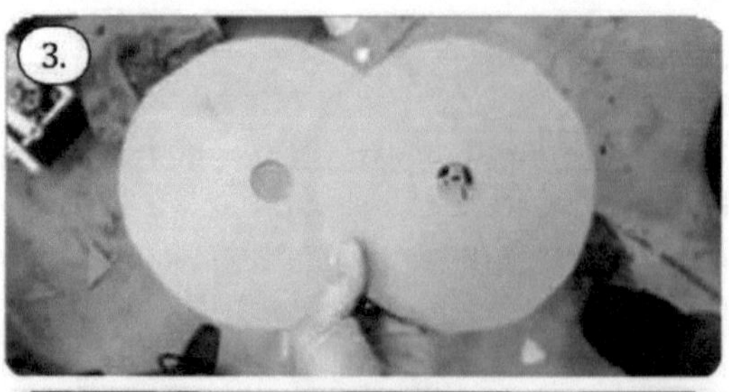

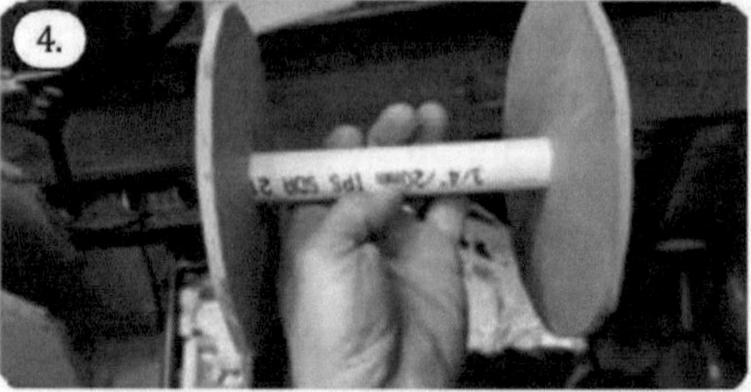

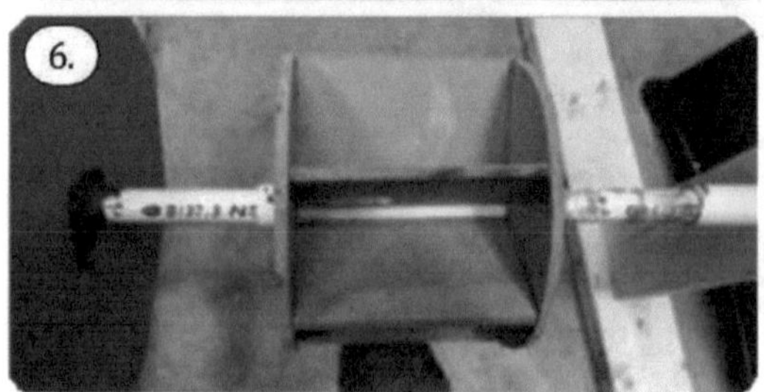

Testing

1. Putting this device into operation depends on where the water source is and how you intend to position the generator. Before placing it in operation, test the generator to make the necessary tweaks and modifications to achieve the voltage you want.
2. I built a test jig out of old 2x4's with a hole in either side to fit a 1" PVC pipe.
3. On one side, attach the stator by sliding it onto the PVC pipe, leaving only enough sticking out to cause the rotor's magnets to be as close as possible to the coils as they spin.
4. Slide the axle into this same pipe so that the rotor's magnets are facing the coils.
5. Slide the other end of the axle into the PVC pipe on the other end of the jig.
6. Connect a multi-meter to the leads on the stator.
7. Set the multi-meter to AC voltage and use a

hose or other water source to spin the water wheel.
8. Check the voltage and make any adjustments as required.

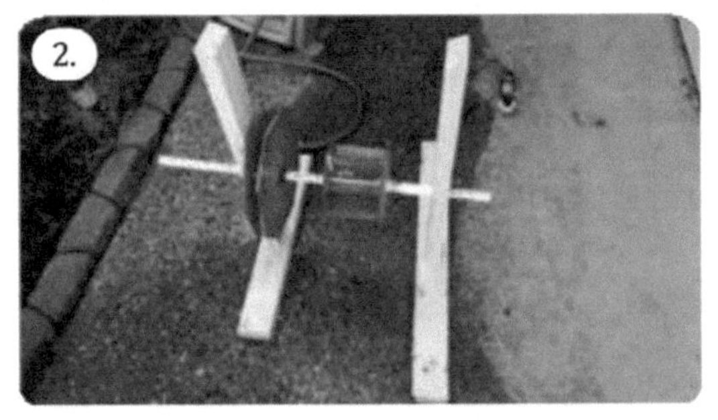

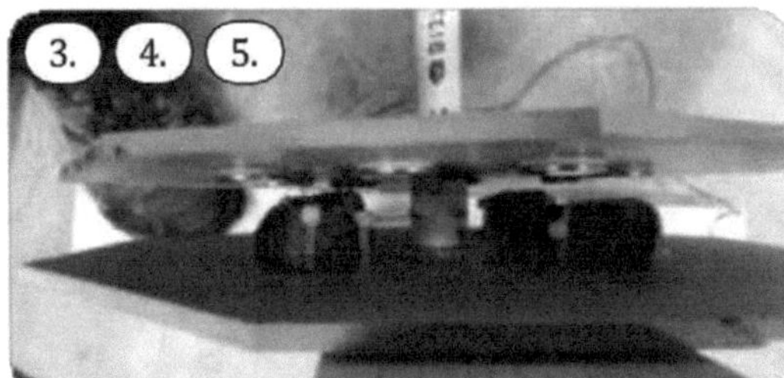

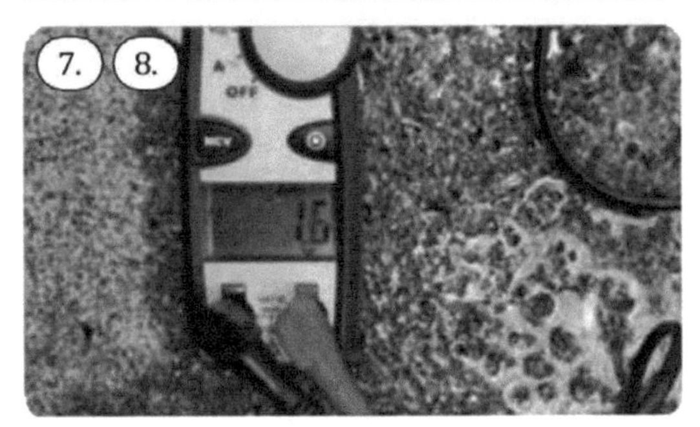

Considerations and Modiflcations

- You will need to use a transformer to switch the current from AC to DC. It is possible that the charge controller that you use to control the charging of your batteries may do this for you.
- Adding coils and magnets will increase the voltage of your generator.
- Consider using bearings and grease to reduce friction, and this will increase the efficiency of the generator.
- Instead of a waterwheel, you could design a turbine or something similar along with directing the water to the best position to achieve the highest RPMs.

While this style of hydroelectric generator will not produce a lot of power, it can be adjusted, modified, and designed to be a constant source of power fed into your battery bank. The knowledge of how this generator works will potentially become very useful in a prolonged grid-down scenario allowing you to design and build your hydroelectric powerplant from materials that will be littered around a post SHTF world.

PROJECTS RELATED TO ELECTRICITY

How To Charge Your Phone When There Is No Electricity

EASY **DIFFICULTY**

What would you do if the power was out and your phone battery was dead, but someone's life would depend on you being able to make a call?

You can charge your phone from another battery. Gone are the days of our grandparents. The modern society we all live in now seems to be unable to function without certain commodities. Among these, electricity seems to be at the top of the list. Electricity is something almost all of us take for granted now. It seems that, no matter what you plan on doing, you will need electricity. Whether you want to clean your house, watch a movie, read a book at night or just charge your phone, you will need electricity.

This chapter will show you how to charge your phone when no power is available - whether you want to prepare yourself in case of a power outage, a storm, or a possible EMP, or you just want to test it for yourself out of curiosity to see if it really works.

Materials Needed

Apart from your phone and its charging cable, you will only need these four items:

- a 9V Alkaline battery
- a spring (you can take one out of a pen)
- a car charger
- some tape

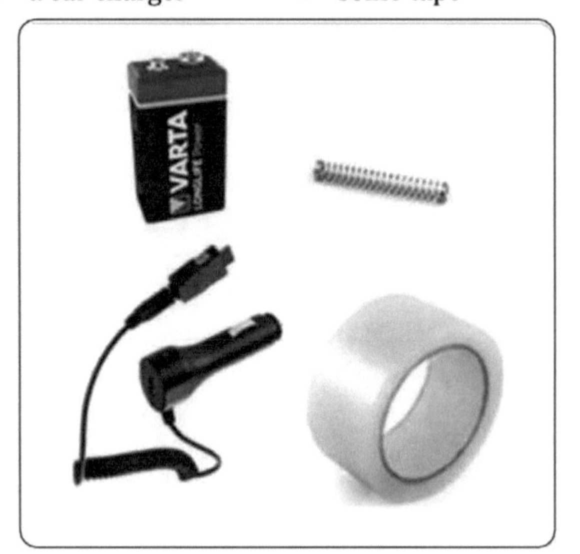

The Step-by-step Process

1. The first step is to take out the spring from a pen. Once it is out, you need to connect it to the negative terminal of the battery (The negative terminal is the bigger one.)
2. The spring will also need to touch the metal part of the car charger's side nub. Use some tape to hold the spring in place, making sure the spring is touching both the negative terminal of the battery as well as the metal nub of the car charger.
3. The next step is to connect the car charger to the positive terminal.
4. Then you need to connect your phone to the car charger using the phone's cable.

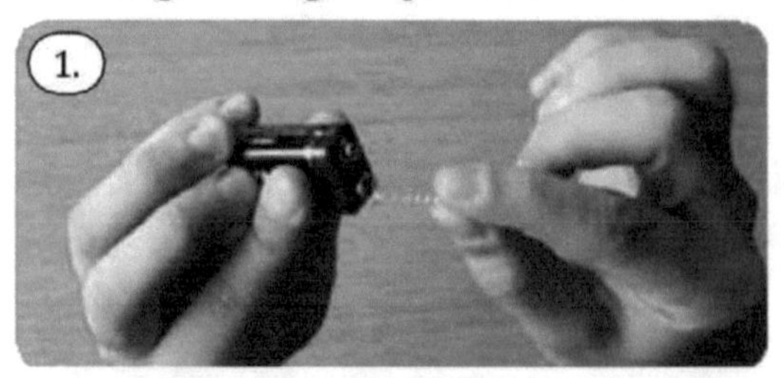

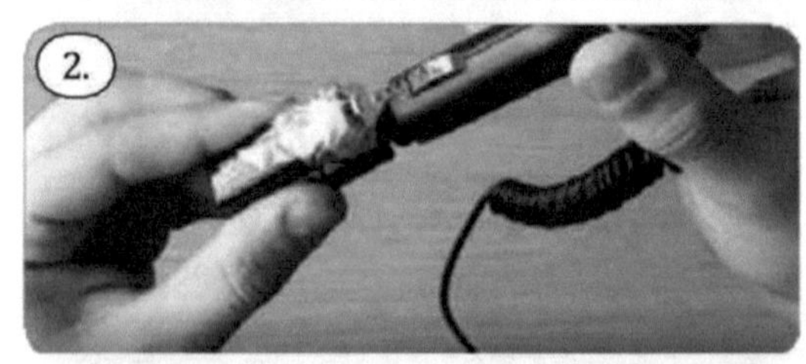

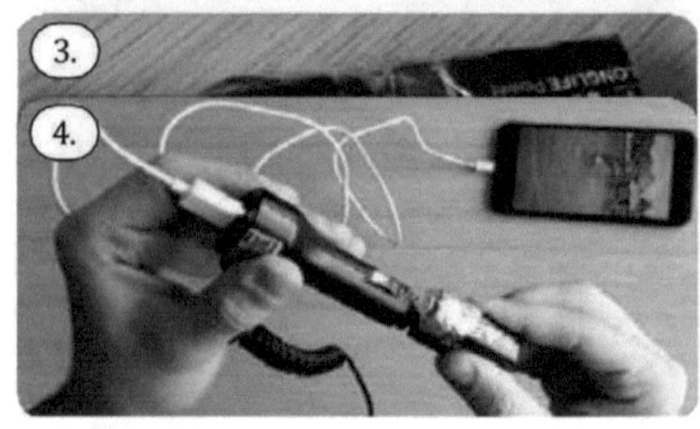

At this moment, the green light of the car charger should be on. This means your phone is being charged.

On average, a 9V battery has about 550 milli-Ampere hours (mAh), while a mobile phone's battery has 2000 mAh. This means your phone's battery will not be fully charged by using only one 9V battery. You would need about four 9V batteries to fully charge your phone, depending on what type of battery you use and what mobile phone you have. But you'll surely make a LOT of phone calls with one battery. Please see the table below for more details:

- Samsung Galaxy S6: 2,550 mAh
- Samsung Galaxy S7: 3000 mAh
- Samsung Galaxy S7 Edge: 3600 mAh
- Samsung Galaxy S8: 3000 mAh
- Apple iPhone 6s: 1715 mAh
- Apple iPhone 7: 1960 mAh
- Apple iPhone 8: 1821 mAh

Who knows, maybe this simple trick can save someone's life by allowing you to make a 911 call.

Type		**IEC name**	**ANSI/ NEDA name**	**Typical capacity in mAh**	**Nominal voltages**
Primary (disposable)	Alkaline	6LR61	1604A	550	9
		6LP3146	1604A	550	9
	Zinc-Carbon	6F22	1604D	400	9
	Lithium		1604LC	1200	9

How To Get Power Out Of Dead Batteries

EASY **DIFFICULTY**

With so many electronic devices in our daily lives, anyone who doesn't have a good stockpile of common sized batteries is asking for trouble. While it is still possible to live without those electronic devices, we've become so accustomed to using them, and the convenience that they provide, that it would be difficult to get by without them, especially in the event of an emergency.

But there are more types of batteries today than ever before. So, between the increased types of batteries and the increased number of devices we all have, it can be challenging having all the batteries we need. If only we could make our batteries last longer; that would save us money, as well as allow us to use our survival devices longer in a crisis.

What if I were to tell you that you could? What if I were to say that "dead batteries" weren't really dead?

Let's talk a little battery theory for a moment. Small batteries fall into two basic categories: rechargeable and single use. Obviously, you can recharge batteries that are designed for it and continue using them. That's not what I'm talking about. What I'm talking about are the batteries designed for single use. We usually throw them away, once they are dead. But what is dead?

For simplicity sake, let's just talk about AA and AAA batteries. These are the most common sizes and whatever applies to them will also apply to all other single use batteries. These battery sizes, as well as C and D cells, are rated at a nominal 1.5 volts DC. I say "nominal" because new batteries actually have a higher charge, typically somewhere around 1.6 volts DC.

The devices we use generally have (use) a number of these batteries, as there is little that actually runs off of 1.5 volts (although there are some devices that do). To get more voltage, the batteries are connected in series. That means that the positive end of one battery (the one with the nipple on it) is connected to the negative end of the next (the flat end). Whenever we do that, the voltage of the batteries is added, increasing the voltage. So, we'll put two batteries in to have a nominal 3 volts or 4 batteries to have a nominal 6 volts.

Batteries produce this electrical power by chemical reaction. As the device is used, it draws power from the batteries, gradually lowering the amount of available chemicals. As this level diminishes, the voltage that the battery produces drops as well.

So, that battery which started out somewhere around 1.6 volts, will keep dropping its voltage until it hits about 1.3 volts. At that point, the device usually stops working. This is the point at which we normally say that the battery is "dead." But it really isn't. The battery just has less power than it needs to have, in order to power our device. But perhaps it can be used for something else.

GETTING MORE OUT OF THE BATTERIES

Okay, so if those batteries still have some power in them, all we need is some way of getting it out. To do that, we're going to do the exact same thing that we did inside the device; we're going to connect the batteries in series. The only difference is, we're going to connect more of them together, than we usually do. By doing this:

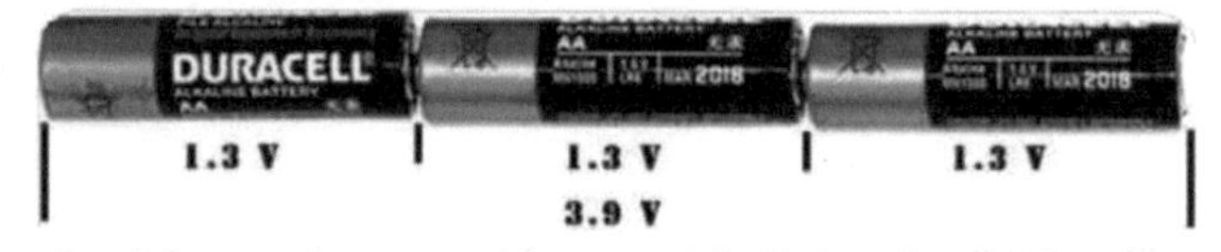

- 4 batteries provide us with 5.2 volts (1.3 x 4)
- 6 batteries provide us with 7.8 volts (1.3 x 6)
- 8 batteries provides us with 10.4 volts (1.3 x 8)
- 12 batteries provides us with 15.6 volts (1.3 x 12)

Obviously, these batteries can't be put in the device this way, because there isn't enough physical space for them inside the device. But many electronic devices have an external power connector, which we could connect to.

Many of today's electronics are designed to be able to be recharged by a computer; connecting it to the USB connector. The USB connector on your computer provides 5 volts. So do the USB chargers that plug into the wall. The main difference is that the ones which plug into the wall will provide a lot more 5 volt power than you can pull out of a USB connector on your computer. That's why some devices won't charge when connected to a computer.
If you look at the battery in many of these devices (cell phones are a great example), it's actually not a 5 volt battery, but a 3.6 to 3.8 volt battery. So, as long as we are providing more than 3.6 volts to the device, it will work. We can get more than that much power out of 4 dead AA batteries.
All we need, is a way of hooking all this up together; some sort of a battery pack that will allow us to connect the batteries together, along with a connector to attach to our electronic devices.

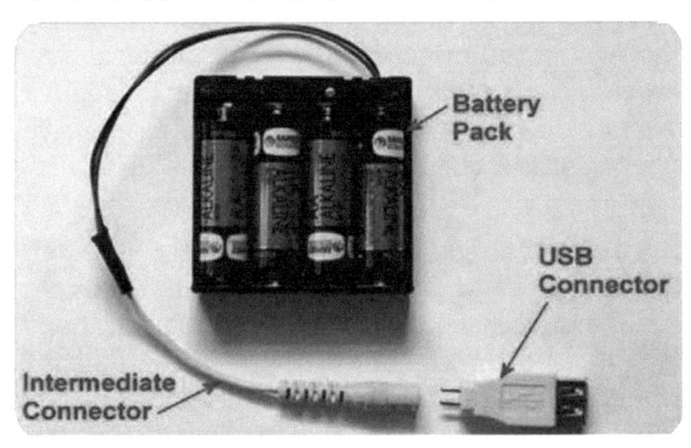

This battery pack holds 4 AA batteries. If we put in 4 dead AA batteries, that means we've got a total of 5.2 (1.3 x 4) volts available to us. There's an intermediate connector that I've attached to it, in order to allow the same battery pack to provide power through a variety of connectors. Finally, since that's enough to run those devices that charge off of a USB connector, I've added a USB connector.
This device will now work to charge or power a cell phone, tablet or digital camera; amongst a host of other devices. All I need is a USB to micro USB adapter, something that is quite common. So, I can continue using those batteries, until their output voltage drops to 0.867 volts each. Then the batteries will have to be replaced.

BUT THEY ARE STILL NOT DEAD

Granted, 0.867 volts doesn't sound like much, but it's still a bit over half the battery's nominal rating. So, we should be able to continue using them a while longer, just by using more batteries connected together in series, than what we had before.
If we connect six of these batteries together in a battery pack, we'll be producing 5.2 volts. If we connect eight of them together, we'll be producing 6.936. Either of these would work for powering those same USB devices. Since battery packs to hold six batteries are not that common, but battery packs for four and eight batteries are fairly common, we might be better off using the eight batteries.
Even though this sounds like too much power for the devices, it's not. All electronic devices have a voltage regulator in them. They have to, as the amount of power the batteries provide isn't constant. We've seen how it goes down. So, the "extra" voltage will be cut off, only allowing the amount of voltage that the device needs to pass through.
In real terms, there are limits to how much voltage we can give a device, before the voltage regulator can't do its job. But that's not a worry here, as we're not going over that limit. Were we to hook 12 volts to a 5 volt device, however, it would probably be too much.
So, there you have it, a great way to save money on batteries, but more importantly, a great way to get more mileage out of your batteries in an emergency survival situation.

How To Make Your Own Solar Panel System

MATERIAL COST  **2.475** **COMPLEX**  **DIFFICULTY** **6 HOURS**

One of the key components to becoming self-reliant is to have the ability to produce electricity outside of the traditional power grid. However, there is a good reason why most people do not exercise energy independence. It is because generating enough renewable energy to power a home is no easy task. However, it is possible to set oneself up with an off-grid power option that will power critical components within the home, such as a refrigerator.

Why Choose Solar?

The Sun is our constant companion during its daily transit across our sky. Since the Sun is guaranteed to rise and set each day and has predictable amounts of daily sunlight, it makes sense that we should convert some of that solar energy into electrical energy.

However, not all areas of the globe are equally suited to use solar power as an off-grid power solution.

Those who live in arid environments can expect to see an unobstructed Sun daily. Therefore, solar power is a very good option since the generous amounts of sunlight will provide maximum output from solar panels. However, suppose we travel further North to above the arctic circle where the Sun's angle is very low with protracted periods of darkness. In that case, solar power can still be used but requires other forms of renewable energy to pick up the slack.

Pros and Cons of Solar Power

Solar power is currently in wide use as an alternative to the traditional means of generating electricity, and like all the other methods, it has its pros and cons.

PROS

- It does not rely on mechanical or moving parts to generate power, which means less maintenance and breakdowns.
- In some areas, electricity can be sold back into the power grid, generating some income to offset the costs of setting up the system.
- Solar panels have a long lifespan.
- Modern systems have become efficient enough to generate electricity on overcast days.

CONS

- Generates no power at night.
- The panels need to be oriented to take full advantage of the Sun's transit across the sky to optimize the system for maximum efficiency.
- Solar panels can be covered in snow or damaged by hail.

Solar power should not be relied on as your only means of off-grid power. Instead, you should have another method of generating enough electricity to recharge your system's battery bank. Ideally, this alternate power should be renewable energy, but a gas or diesel generator is acceptable.

SAFETY

Before continuing, you need to educate yourself about the risks of setting up a solar power system, like the one I will describe here. The amperages involved throughout the system can deliver electrical shocks, which have the potential to be fatal. Exercise extreme caution in all aspects of installing and using this system, and make yourself aware of all the rules, regulations, and building codes in your area before constructing your solar power system. Also, exercise caution in all aspects of installing and using this system, and make yourself aware of all the rules, regulations, and building codes in your area before constructing your solar power system. Also, exercise extreme caution when installing panels on your roof, as a fall can result in serious injury or death.

Basic Electrical Terminology

This chapter will use a lot of technical jargon when describing the components of the solar power system. A basic overview of these terms and what they represent is useful at this point.

Volt (V) - Voltage is the measure of electrical force to produce current.

Amp (A) - An ampere is the measure of the flow of current through the conductor.

Watt (W) - A watt is a way we measure electrical

power. To find the wattage, we only need to multiply the Volts by Amps.
Ampere-Hour (Ah) - A unit of measurement used to represent battery capacity. For example, a 10Ah battery will provide 10A of power for one hour.
Watt-hours (Wh) - A watt-hour is the number of watts of energy used in one hour.
Direct Current (DC) - Direct current is an electric current that flows in one direction only. DC power is often associated with electrical circuits powered by batteries.
Alternating Current (AC) - Is an electrical current that reverses direction multiple times a second. Most household appliances operate on AC power.

Essential Components of a Solar Power System

Solar power systems can become very complex, but when we break it down to the basics, they all have a few basic components: solar panels, a charge controller, an inverter, and a bank of batteries.

Batteries

The battery bank is the heart and soul of your solar power system, along with probably being the most expensive component. Batteries come in a wide variety of voltages, but since 12 volts is the most common voltage for a small solar power system, I will be using this voltage throughout this chapter. Of course, there are many reasons you would want a higher or lower voltage battery bank, but to keep this instructional as basic as possible, I will stick to describing a simple 12-volt solar power system. There are many battery choices, but I will speak to the most commonly used battery types and their strengths and weaknesses.

Lead Acid Batteries

Lead-acid batteries are a popular choice for solar battery banks mainly because they are more affordable than lithium. Lead-acid batteries are also the oldest form of rechargeable battery, and chances are there is a lead-acid battery under the hood of your car.

Lead-acid batteries come in two main types: Flooded Lead Acid (FLA) and Absorbent Glass Mat (AGM).

a) FLA Batteries

FLA batteries are filled with liquid electrolytes that are responsible for storing and delivering electricity. This means that FLA batteries need to be stored upright and require good ventilation since they release toxic gasses. FLA batteries also require regular maintenance and are large and heavy. They are, however, cheaper than both lithium and AGM batteries.

b) AGM Batteries

AGM batteries, on the other hand, have fibreglass inside of them to suspend the electrolytes. Storing the electrolyte in this way means that the batteries do not have to remain upright. Also, they require no maintenance and do not release any toxic gasses.
Regardless of the type of lead-acid battery, **DO NOT** discharge them below 50 percent of their capacity. This means a lead-acid battery bank that is 600ah will only have 300ah of usable capacity, so keep this in mind when designing your system.

Lithium Batteries

Lithium batteries are available in many types, but the one most commonly used in solar power systems is Lithium Iron Phosphate (LiFePO4). These batteries are far lighter and smaller than SLA batteries, as well as being more efficient. These batteries have a few advantages over their SLA counterparts, most notably that they are safer and more stable. In addition, lithium batteries are designed to be discharged much deeper than SLA batteries, up to 80% of their capacity.
One significant disadvantage is that lithium batteries are far more expensive than SLA batteries, which often financially put them out of reach.

Solar Panels

Eventually, your battery bank will need a recharge, and in a grid-down situation, you will need to seek renewable power sources. For example, solar panels harness the Sun's energy and convert it into electric current through the photovoltaic effect. What happens is that photons from the Sun get absorbed by the solar cells in the panel, which caus- es electrons to excite and become free. These free electrons are what creates the electrical current. Solar panels come in various shapes and sizes. When we talk about the size of a solar panel, we usually refer to the number of watts they produce under full and direct sunlight. Solar panels can be flexible, with some models which can be rolled up or folded.

Charge Controller

A charge controller does more than regulating the

flow of current from your panels to the battery bank. They prevent the battery bank from both o- ver and undercharging. They also distribute pow- er to any DC loads that you have connected to the charge controller. This component is one of the most important to install since it protects the most costly part of your solar power system.

Charge controllers come in two types: Pulse Width Modulation (PWM) and Maximum Power Point Tracking (MPPT).

PWM controllers turn on and off rapidly, acting as a throttle for the power coming from the solar panels to the batteries. These charge controllers are an older technology and require that the nom- inal voltage of the solar panels match the voltage of the batteries. As a result, PWM charge control- lers are less efficient than their MPPT counter- parts, but they are often better for smaller solar power projects since they are less expensive.

MPPT controllers can be up to 30% more efficient than PWM controllers and can handle higher voltages from solar panels. These charge controllers will continuously monitor the power levels generated by the panel and automatically adjust to find the best combination of voltage and cur- rent to deliver the most power to the batteries.

Inverter

Solar power systems produce direct current (DC), which is incompatible with household appliances that use alternating current (AC). The device that we use to convert DC to AC is an inverter. Manu- facturers classify their inverters based upon the number of output watts. When selecting an invert- er, we need to choose one that can produce great- er wattage than the number of watts we expect our appliances to consume. It is important to buy an inverter that delivers enough power to handle the momentary power surge when some devices start; in the case of a fridge, that surge of power can be 1500 Watts!

Inverters are classified not only by their output wattage but also classified by how they convert power. Most inverters that you will find are going to be either pure sine wave or modified sine wave. **Pure Sine Wave** inverters produce a waveform that most closely resembles the grid's AC power. Pure sine wave inverters are needed if you want to tie your solar power system into grid power or if you plan on running complex electronics from solar energy. They are also preferred for larger systems.

Modified Sine Wave inverters don't produce the clean sine wave that a pure sine wave inverter does. Instead, the waveform is more choppy, and while this is ok for some devices, any complex or delicate electronics should not be connected to a modified sine wave inverter.

Battery Monitor

Battery monitors are not required, but they can help you track and monitor the usage and state of your battery bank. In addition, they can help you to get a better understanding of the capacity left in your battery bank instead of simply monitoring the voltage and making an educated guess as to how close to 50% of capacity your SLA batteries are. Try to find charge controllers of inverters that include a battery monitor.

Cables and Wiring

To tie all of this together will be a mess of cables and wires, with every wire and connection being a possible failure point. It is very important to select the proper wire size for each part of the system and ensure that the appropriate connectors are in the correct places.

- Wires that run from the solar panels to the charge controllers need to be water-resistant and usually come with an MC-4 connector which forms a water-tight seal when plugged together.
- The battery cable needs to be rated for the amount of power you plan to run through them. Therefore, it is good to oversize the cables and follow the manufacturer's recommendations when purchasing them.
- It is also advisable to install a fuse or breaker on the positive cable that leads from the positive battery terminal to the inverter. Doing this will prevent the inverter from trying to draw too much amperage from the batteries.

Designing a Solar Power System

Designing a solar power system can be a very complex process, but it need not be. As long as you take the process step by step, you will be able to determine how your system will look with relative ease.

Start by answering a few questions:

1. WHAT ARE YOU GOING TO POWER?

The appliances that you are going to power with your solar power system is the primary driving factor in the design of your system. So, to begin, write a list of all the appliances you want to run from your solar power system.

For Example:

- Refrigerator
- Lights
- Charge laptop and phones

Once you have a list, make a table including the Watts the appliance uses on average. Include a column for the number of hours in a day that the device will be used.

For Example:

APPLIANCE	WATTS	HOURS OF USE
Fridge	100	24
Lights	60	5
Laptop	8	2
Phone	6	2
Microwave	1000	.5
Freezer	50W	24

Add a fourth column to multiply the wattage and hours for each appliance giving you the Wh. Add this column to determine the daily Wh for the appliances you want to run in a grid-down scenario.

For Example:

APPLIANCE	WATTS	HOURS OF USE	WATT-HOURS
Fridge	100	24	2400
Lights	60	5	300
Laptop	8	2	16
Phone	6	2	12
Microwave	1000	.5	500
Freezer	50W	24	1200
TOTAL Wh			4428

As you can see from the above table, this example gives us a daily Wh of 4428, which we could round up to be 4500Wh. So that means that your solar panels need to deliver a minimum of 4500Wh of power over a day to replace the energy you used in the battery bank. This does not consider the inefficiency of the inverter or that some days will produce less power than others. In reality, your solar panels will need to create an excess of power to charge your batteries sufficiently.

2. HOW MANY PANELS DO YOU NEED?

Before calculating how many panels you need to deliver that much power, you first need to determine how many hours of peak sunlight your panels are likely to receive. Fortunately, there are many online resources that you can use to find out how many hours of peak sunlight you can expect. One such resource is this:
unboundsolar.com/solar-information/sun-hours-us-map

Let us say, as an example, your area receives 5 hours of peak sunlight a day. That means, in five hours, your panels need to produce at least 4500 watts of electricity.

Let's assume that you are going to use 200W solar panels, and they are optimally placed. Then, all we need to do is multiply the panel's wattage by the hours of peak sunlight. So, if your location gets 5 hours of peak sunlight a day, we can determine how much power each panel will provide during a day.

For Example:

200W Panel x **5** Hours Peak Sunlight = **1000W** of daily power.

Our example appliance's daily power consump- tion would require at least five 200W solar panels! This chapter will centre around building a system to power only a small refrigerator for 24 hours a day instead of a list of appliances. This will pro- vide a basic and clear picture of the steps involved and give you, the reader, the foundation to build your system.

If we select an average household fridge, we can expect average power consumption of 100W to 400W depending on the size, how often the doors are opened, how much thermal mass it is cooling, etc. In this example, I will use the low end of the wattage scale to demonstrate what a system that achieves the bare minimum would look like.

Once we do the math, we can determine what the average daily Wh for our fridge is.

100W x **24** hours = **2400Wh** a day

In this example, we only need three 200W panels to replenish the power that our refrigerator uses with 600W of power to spare, as long as we get five hours of peak sunlight each day. The extra ca- pacity will come in handy for days where clouds obscure the Sun or that you need some extra ener- gy to charge devices.

3. **HOW BIG OF A BATTERY BANK DO YOU NEED?**

Now that you know how to size your solar panels, we need to figure out how big a battery bank we need to run our fridge. One question we need to ask first off is:

4. **HOW MANY DAYS DO WE WANT THE BATTERY TO PROVIDE POWER WITHOUT CHARGING?**

There may be cases where your solar panels are not producing enough current to charge the bat- tery bank. We need to account for this when de- signing our system. While it is a good idea to have a battery bank that will power your appliance for a week without needing a charge, this will result in a large and expensive battery bank. Being pre-pared for an extra day without the ability to charge the bank is a good start.

We know that our fridge will use 2400Wh a day, and if we want two days of autonomy, we will have a total power requirement of 4800Wh. These fig- ures are AC power though, what we need is to know how much DC power we need.

5. **HOW DO WE FIGURE OUT HOW MUCH DC POWER WE NEED TO PROVIDE THE RE- QUIRED AC POWER?**

Next, we need to know what our inverter ineffi- ciency is. When an inverter converts DC to AC power, some of that power is lost in the process. We need to account for this when calculating the power the battery needs to provide the inverter. For example, if your inverter is 90% efficient, you need to add 10% to your AC watt-hours to find the DC watt-hours.

Example:

10% of 4800Wh is 480Wh, so we need to add these together for our required DC input power to the inverter.

4800 + 480 = 5280Wh of DC power

6. **HOW DO WE CONVERT WH INTO AH TO PICK THE RIGHT BATTERIES?**

Now that we have arrived at a figure for the a- mount of power our battery bank needs to deliv- er, we must figure out how big that battery bank needs to be. Since batteries are measured in Amp- hours (Ah), not Watt-hours (Wh), we need to do some math to convert these values.

The formula for converting Wh to Ah is either:

Wh=Ah*V or **Ah=Wh/V**

If our system is going to be a 12V system, we can plug the values into the equations like this:

Ah=5280Wh/12V

This gives us **440Ah** as our battery size for two days of autonomy from needing to be recharged. If you are like me and have alternate plans for charging batteries without the Sun and want the battery bank to be only enough to power the fridge for 24 hours a day with the batteries recharging every day, then the battery bank size would be **220Ah**.

The longer you want the batteries to power your appliance without recharging, the larger the sys- tem's capacity. So in this example, the battery bank capacities in relation to days with no recharging would look like this.

Zero days - **220Ah** One day - **440Ah** Two days - **660Ah** Three days - **880Ah**

Four days - **1100Ah** Five days - **1320Ah** Six days - **1540Ah** Seven days - **1760Ah**

As you can see, a battery bank that will sustain your fridge for a week of no sun would be very large and expensive.

7. **HOW MANY BATTERIES TO BUY?**

Now that we have the number of Ah required to power a fridge 24/7, we need to buy and install batteries for the battery bank. On the less expen- sive side are the SLA batteries, which are larger and heavier than lithium batteries. However, SLA batteries can not be discharged below 50%, which means that the total capacity of the battery bank needs to be double the required Ah to account for this.

On the other hand, lithium batteries can be discharged much deeper, so the total capacity of the battery bank can be closer to the required Ah of the system.

In this example project, I selected deep-cycle AGM batteries, which are 100Ah each. The six batteries of my battery bank, when connected in parallel, give me a voltage of 12V and 600Ah of total capacity or 300Ah of usable capacity.

The number and capacities of batteries you buy will be largely determined by the availability of the batteries and the space you have available to store them.

8. HOW DO I WIRE THE BATTERIES TOGETHER?

There are two ways that batteries are wired together, and each of these methods gives a different result.

Series - Wiring a battery bank in series means that the first battery has its negative terminal connected to the positive terminal of the next battery and the positive terminal connected to the negative terminal of the next battery. This is repeated throughout the battery bank. Doing this increases the overall voltage of the bank while keeping the capacity in Ah the same.

Parallel - Wiring a battery bank in parallel means that all the positive terminals are connected, and the negative terminals are connected. What this does is increase the overall capacity of the battery bank while keeping the voltage the same.

Building the Battery Bank

The battery bank is the most expensive part of your solar power system. Therefore, it is important to store the batteries in an area where they are kept safe from damage and do not pose a safety hazard to others in your home. One way to secure a battery bank is to build a shelf or box for them. In my case, I wanted to build a rolling battery box that could also hold the charge controller and inverter. It would be good to roll the batteries from one location to another without removing all the batteries, only to set them up again in another area.

Tools Required

- Drill with bits and drivers
- Tape measure and square
- Saw

Materials

- Batteries - I used Sealed Lead Acid Batteries similar to these that are available on Amazon for $ 274.99 USD each

www.amazon.com/dp/B00YB26RYG

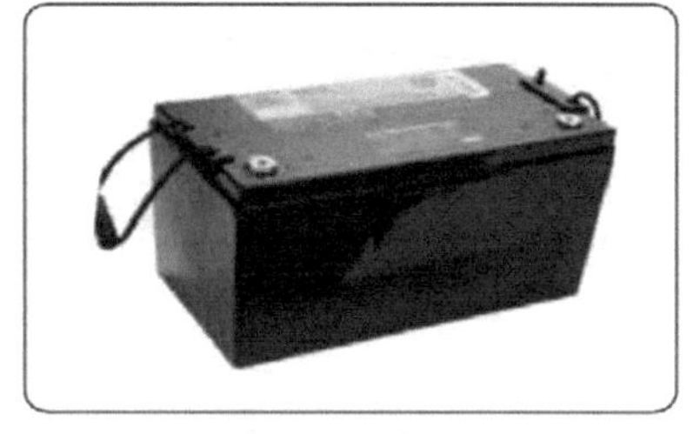

- Battery Cable - What I used was similar to this available on Amazon for $ 152.97 USD and includes cable lugs

www.amazon.com/Welding-Battery-Flexible-Terminal-Connectors/dp/B01MD1YL1I

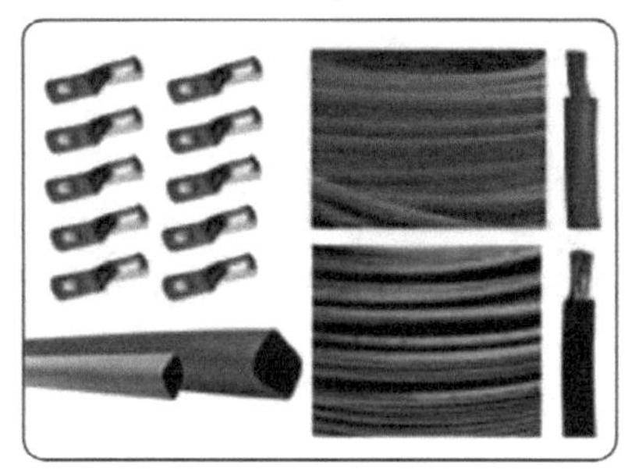

- 2 x 4 Lumber - Available at Home Depot for $ 8.37 USD each

www.homedepot.com/p/2-in-x-4-in-x-8-ft-Premium-Kiln-Dried-Whitewood-Framing-Stud-Lumber-96022/315592380

- Casters - These are available from Amazon for $22.99USD for a set of four

www.amazon.com/Casters-Locking-Polyurethane-Castors-HARDWARE/dp/B08HS381FH

- Charge Controller - This is the one I used which is $ 69.99 USD on Amazon

www.amazon.com/Controller-Discharge-Regulator-Protection-Anti-Fall/dp/B083KN9H22/

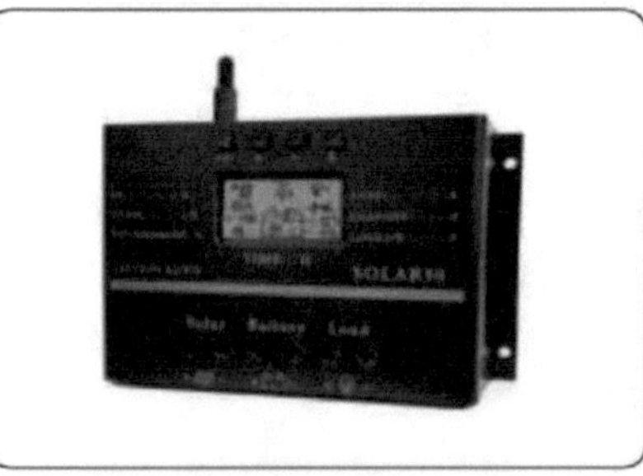

- Inverter – I used this one which is $ 189.99 USD on Amazon

www.amazon.com/Invert-er-2000Watt-120Volt-Control-Display/dp/B07R511SR8

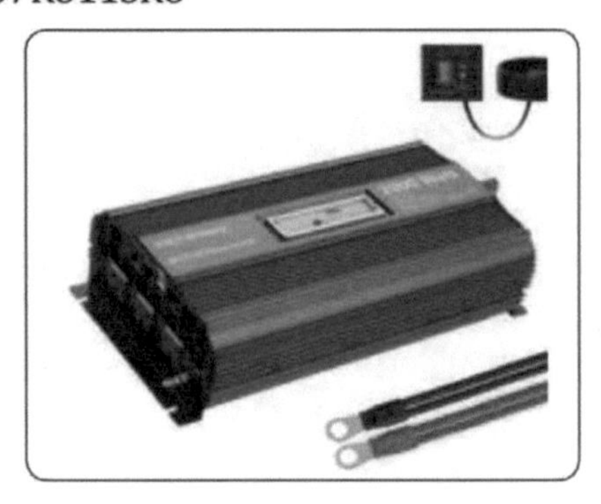

Construction of the Battery Bank

1. Begin by measuring your batteries and figuring out what configuration you want the batteries to be. I laid the batteries out in a 2 x 3 configuration with a little space between each battery. Suppose you have a significant amount of batteries; you may not want to construct the rolling unit that I am. Instead, build a basic shelf and do not store the batteries on bare concrete.
2. Once I had my overall size, I cut material to make a basic frame for the bottom of my battery bank.
3. Once cut, I screwed all the pieces together, making the base of my battery bank.
4. I flipped the frame over and screwed the casters in place.
5. I loaded all the batteries onto the frame.
6. Once all the batteries were in their final orientation, I measured the distances between the terminals of the batteries to determine the cut size for my battery cables.
7. I cut the appropriate number and lengths of cables to connect the terminals in a parallel configuration.
8. I then crimped the lugs to each end of the cables.
9. Once the cables were put together, I wired the battery bank in parallel. I made sure to connect the positive terminals first then the negative terminals.
10. Once the battery bank is wired, I measured the overall height for a shelf to mount the inverter and charge controller.
11. I then measured my inverter so I could cut lumber to build the shelf to mount it.
12. I cut lumber to make four uprights that would hold up my shelf. I also cut two cross members to complete this shelf. I then attached these pieces to the battery bank.
13. I mounted both the inverter and the charge controller to the shelf.
14. Finally, I wired the batteries to the charge controller and inverter. I first connected the positive cable from the positive terminal on the charge controller to the first positive battery terminal. Then, I also wired the positive cable from the inverter to the same positive battery terminal.
15. I connected the negative cable from the inverter to the negative terminal on the same battery.
16. I connected the negative cable from the charge controller to the negative terminal on the last battery in the battery bank. This allows for more even charging of the battery bank than if I connected the positive and negative cables to the same battery in the system.

5.

6.

7.

8.

9.

10.

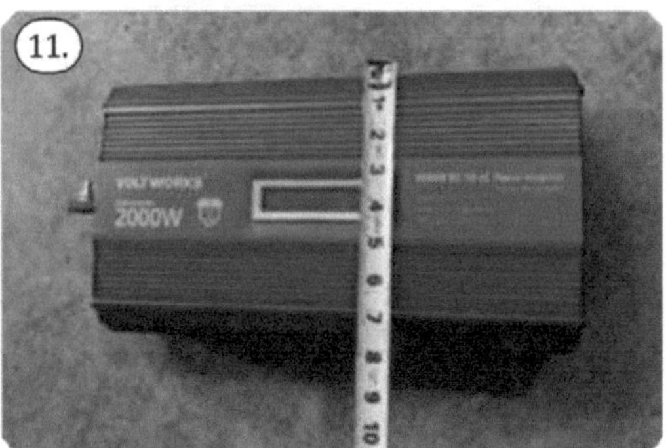
11.

12.

13.

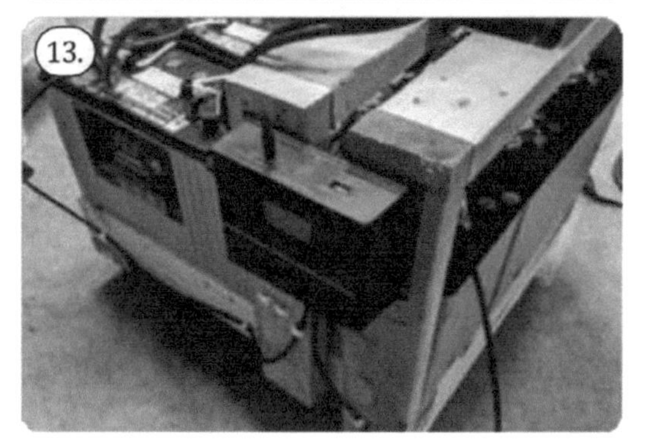
13.

14.

Considerations

There are several considerations when putting a battery bank together.

- First is the type of batteries in the system. Flooded lead-acid batteries need to be stored upright and have off-gassing and maintenance concerns, while AGM batteries can be stored in any position and don't produce dangerous gases. Lithium batteries, on the other hand, are considered to be very safe and stable. Always check the manufacturer's instructions and safety guidelines before building your battery bank.
- Always connect the positive cable first and always disconnect the negative cable first.
- Keep the inverter and charge controller close to the battery bank to cut down on the power loss that will result from long runs of wire.
- Store the battery bank in an area with no flammable materials in case of sparks igniting any combustible gases.

Installation and Wiring of the Solar Panels

There are a lot of methods and places that a solar panel can be mounted. The problem is that there are so many different roofing materials and styles that it is impossible to demonstrate how to mount the panels to a roof adequately. That being said, the positioning, wiring, and considerations are universal and will be described in detail here.

Deployment of Solar Panels

There are a couple of ways that we can deploy solar panels. First, we can permanently mount them to our roofs to catch the most sunlight over a day. There are also motorized mounts that solar panels can be mounted on, which track the Sun's course through the sky. Unfortunately, these mounts are very expensive and probably out of reach financially for most people.

Permanently mounting a solar panel is a great option because it allows for a constant power flow to the charge controller. The only downside of this setup is that if you only intend to use the power system for emergencies, it may not be practical to have permanent panels on your roof, risking damage.

Solar panels can also be deployed when needed and stored inside when not required. This gives you the flexibility to deploy either full-sized solar panels or smaller folding solar panels. It also gives you the ability to alter the angle and position of the panels to make the best use of the available sunlight.

Ideal Solar Panel Placement

Try to place solar panels to expose them to a maximum amount of direct sunlight to deliver the highest power levels to your battery bank. Unfortunately, since most of us could not dream of affording solar panels that track the Sun through the sky, we will have to do the best with what we have. In the Northern Hemisphere, solar panels should face True South, and the opposite is true for the Southern Hemisphere. This is because the Earth's tilt places the Sun to the South in the Northern Hemisphere and vice versa. Placing the panels in this way is the best way to maximize their exposure to the Sun's rays.

To best catch the Sun's energy, it is recommended to tilt the solar panels as well. As a rule of thumb, the tilt angle will equal the latitude where the panels are situated. So, for example, if you live at 50 degrees latitude, you should tilt your panels so they are at 50 degrees.

Realistic Solar Panel Placement

Most of us will place solar panels on the roof of our homes or have panels that we can deploy rapidly in an emergency. However, as far as mounting panels to the roof goes, you are limited by which direction your roof faces and the pitch.

There are options for tilting roof-mounted panels

to achieve the ideal angle, but if you may not have areas of your roof that face the appropriate direction. You can either place panels on the ground or position roof panels exposing them to whatever sunlight you can. Less than ideal conditions may result in adding many more solar panels to your array to make up for the inefficiencies.

Wiring Solar Panels

How you wire your solar panels has a lot to do with what charge controller you have. MPPT charge controllers are designed with panels wired in series, while PWM charge controllers pair with panels connected in parallel.

Wiring our panels in series is similar to what happens when we wire batteries in series. The voltage increases, but the current capacity stays the same. In comparison, wiring our panels in parallel keep the voltage the same but increases the current capacity.

When we wire panels in series, the whole system will fail if something happens to one panel. So, for example, if one panel were under some shade, it would drag the rest of the system down with it.

Panels wired in parallel do not suffer this fate. Instead, if one panel fails, the current will bypass the failed panel.

Mounting Panels on the Roof

Do not attempt to fabricate brackets or mounts for your solar panels. Instead, purchase installation brackets designed for the style and type of roofing material that you have.

Some examples of mounting systems are:

- Here is an example of a bracket that can adjust to achieve the proper tilt and costs $ 39.99 USD:

www.amazon.com/Adjustable-Brackets-support-Surface-off-grid/dp/B00IYWBOLA/

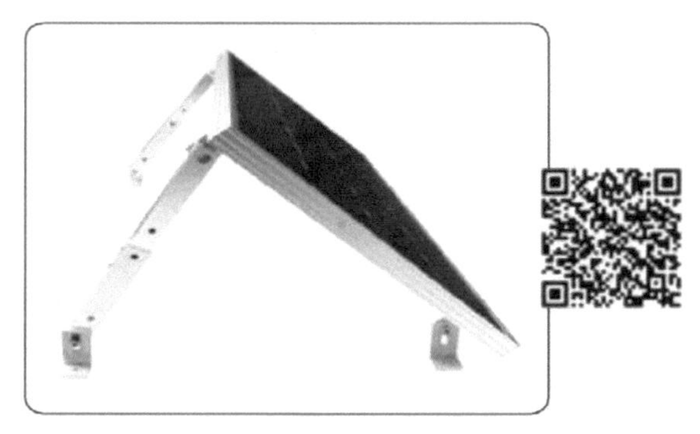

- These are very simple brackets that cost $ 11.39 USD but don't allow for tilting:

www.amazon.com/Renogy-Solar-Mounting-Bracket-Supporting/dp/B00BR3KFKE/

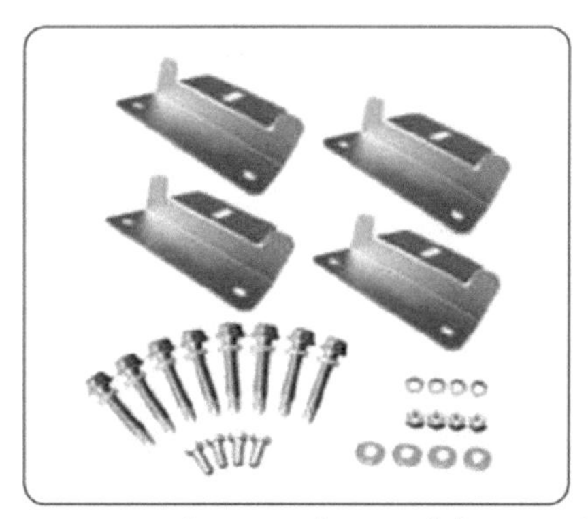

Always exercise caution and keep safety in mind when installing panels on your home's roof. If there is any doubt about your ability to complete the task effectively, it is best to hire professionals to do it for you.

DIY Ground Mounting Option

It may be the best option to install your panels on ground level, in which case you will need to build some stands to hold the panels at the appropriate angle. These are very easy to construct and only require some easily obtained lumber and screws.

MATERIALS

- 200W Solar Panels – I used these which are sold on Amazon for $ 184.99 USD a pair

www.amazon.com/WEIZE-Watt-Monocrystalline-Solar-Panel/dp/B08721CBST/

- Heavy gauge wire with MC-4 Connectors – What I used was similar to these, which are

available on Amazon for $ 144.99 USD
www.amazon.com/WindyNation-Gauge-Black-Extension-Connector/dp/B01D7VBT2G/

- MC-4 'Y' Cables – Like these that are sold on Amazon for $ 8.49 USD

www.amazon.com/PowMr-Branch-Connectors-Parallel-Connection/dp/B0822QMRCW/

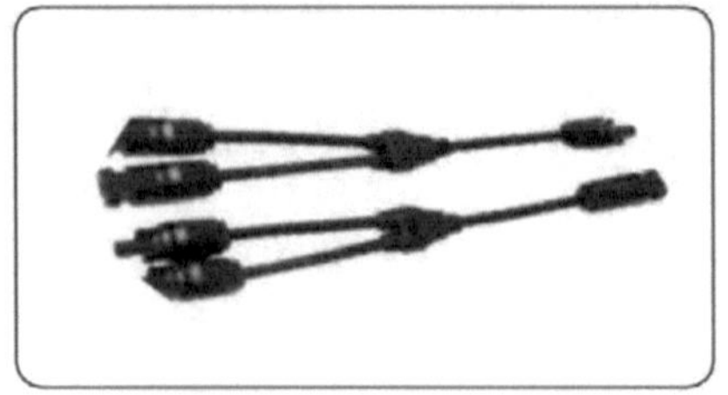

- 2 x 4 lumber - Available at Home Depot for $ 8.37 USD each

www.homedepot.com/p/2-in-x-4-in-x-8-ft-Premium-Kiln-Dried-Whitewood-Framing-Stud-Lumber-96022/315592380

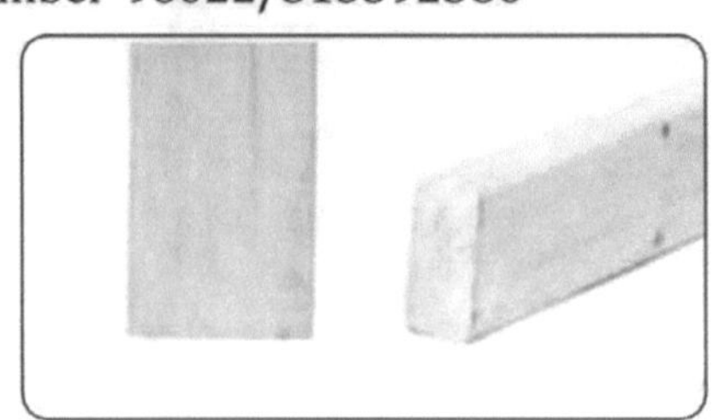

INSTRUCTIONS

1. Measure the length and width of your solar panel.
2. Cut lumber to build the frame on which the panel will rest. Make sure this frame will go at least halfway up the panel. Assemble these pieces by screwing them together.
3. Lean the frame against a wall and use an angle finder to position it for the appropriate amount of tilt.
4. Lay a 2x4 on the floor alongside your frame. Butt the end of the 2x4 to the wall and scribe a line where the frame and this 2x4 intersect. Cut along this line and use this piece as a template for all the bottom side supports.
5. Cut lumber to build the frame on which the panel will rest. Make sure this frame will go at least halfway up the panel. Assemble these pieces by screwing them together.
6. Lean the frame against a wall and use an angle finder to position it for the appropriate amount of tilt.
7. Cut and install a cross-member to add additional support.
8. Install the solar panel.

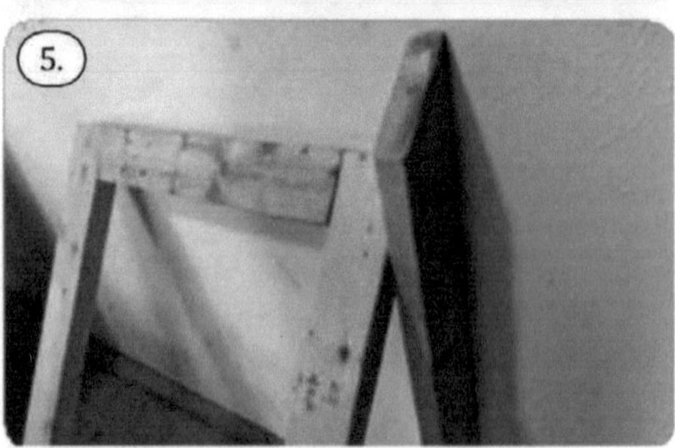

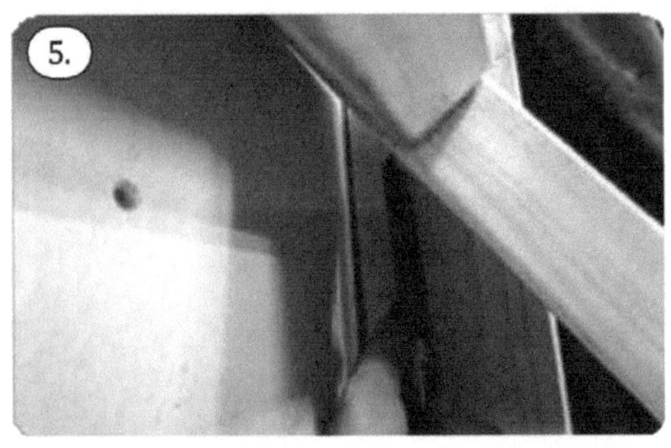

Wiring the Panels in Parallel

In my case, I want to connect my panels in parallel because the charge controller that I selected is PWM and not MPPT.

When connecting wires to panels and panels to charge controllers, **ALWAYS COVER THE PANELS**, so they do not generate any current when making the connections.

1. Start by identifying which connector on the solar panel is negative and which is positive.
2. Connect the positive leads by using a 'Y' connector or a coupler.
3. Connect the negative leads in the same way.
4. Next, take a length of extension cable with a male connector on one end and a female connector on the other end that is twice the distance from your solar panel to the charge controller. We need the length to be twice the size required because what we want to do is cut this cable in half to get one positive and one negative cable.
5. Find the halfway point.
6. Cut and strip the cable.
7. Mark the positive and negative ends, then connect to the charge controller.
8. Connect the positive end to the positive side of the panels, then connect the negative end to the negative side of the panels.

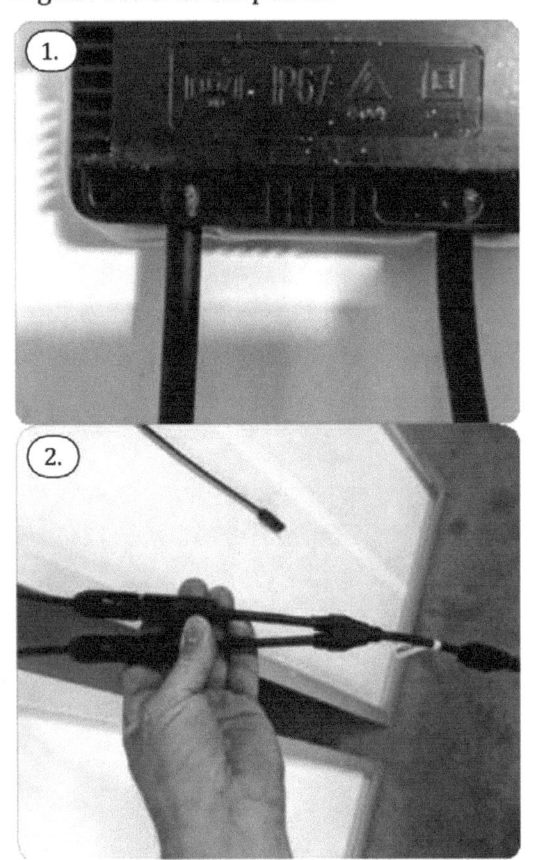

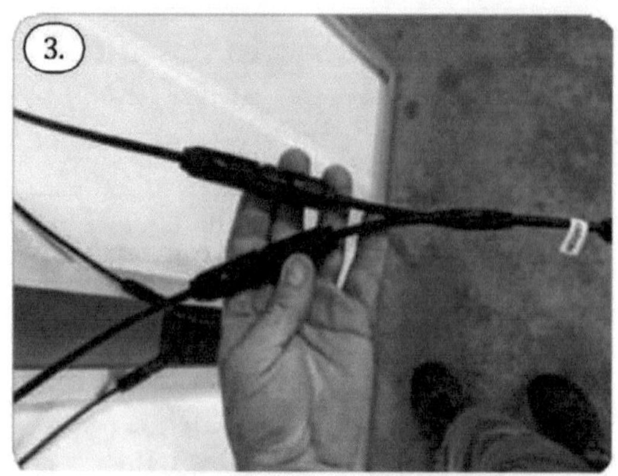

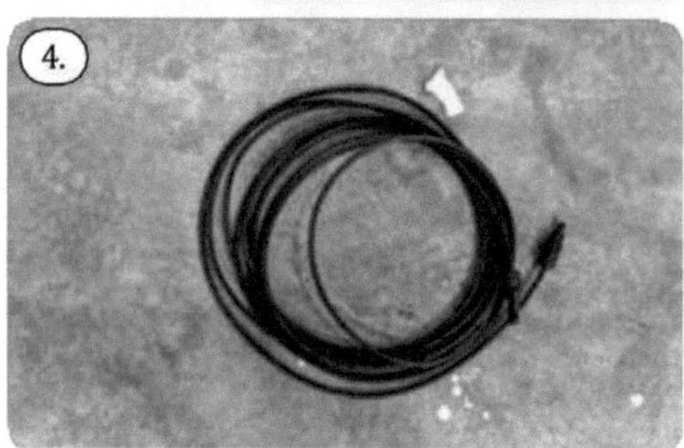

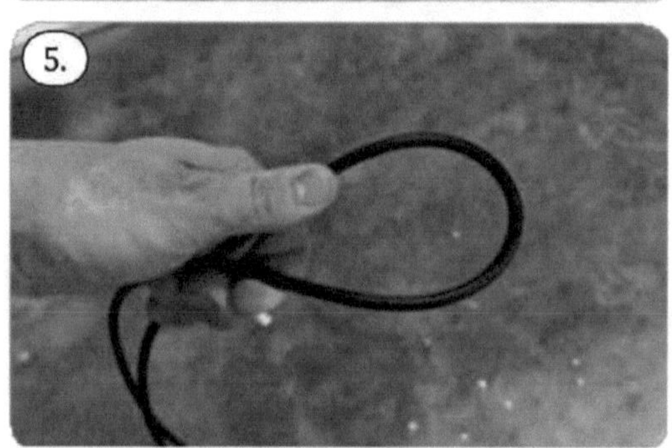

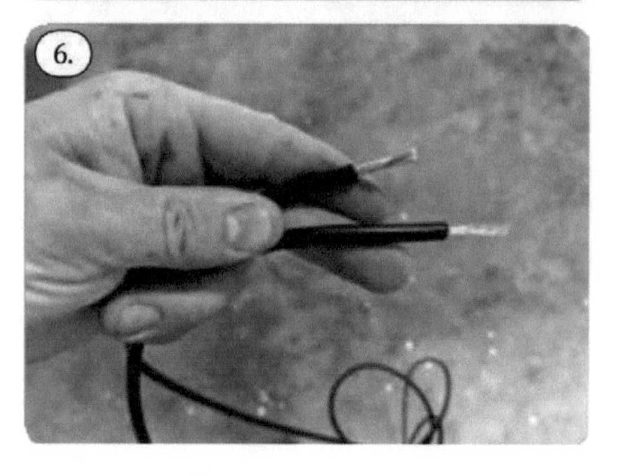

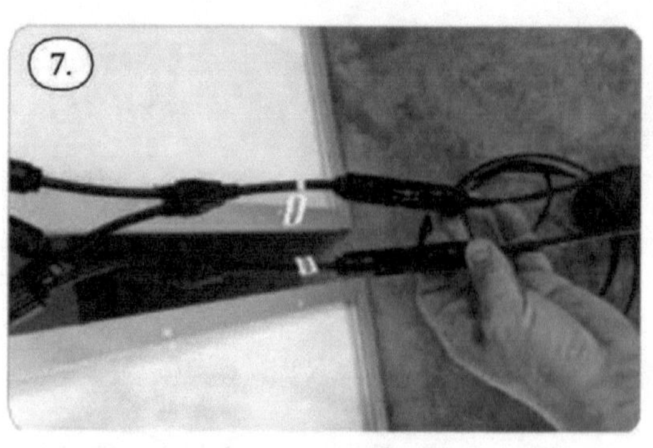

Wiring the Panels in Series

There are cases where we want to have solar panels wired in series rather than parallel. To connect multiple panels in series, start by covering the panels to avoid electrical shock.

1. Identify the positive and negative ends.
2. Starting with the first panel, connect the negative power cable to the positive on the next panel.
3. Repeat this process with each panel.
4. At one end of your row of panels will be a positive connector, and the other end will be a negative connector. As with connecting panels in parallel, cut an extension cable in half to get two lines that are the lengths that you need to reach from the panels to the battery bank.
5. Connect the extension cables to the charge controller.
6. Connect the positive end of the cable from the charge controller to the positive side of the panels.
7. Connect the negative cable to the negative end of the panels.

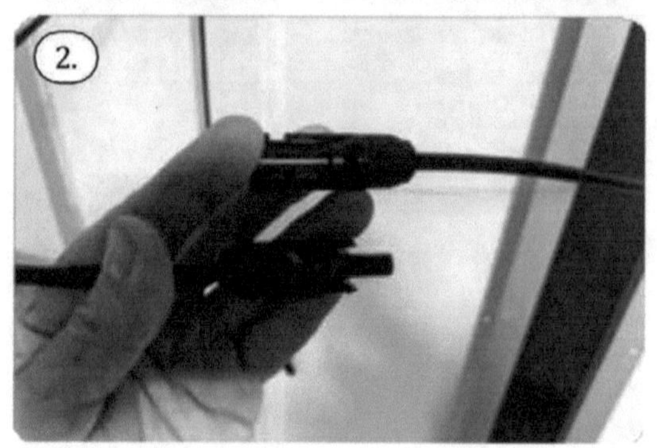

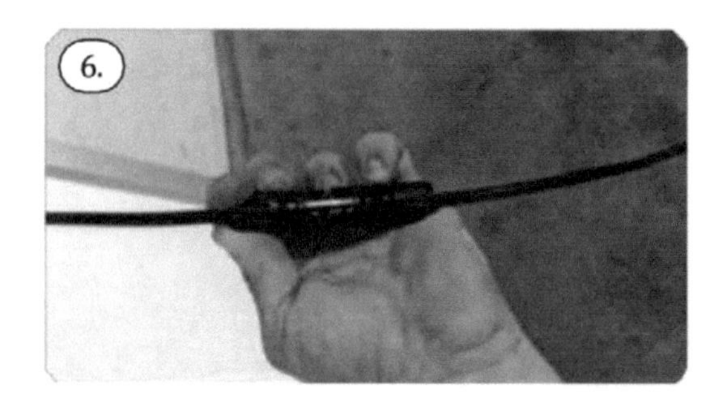

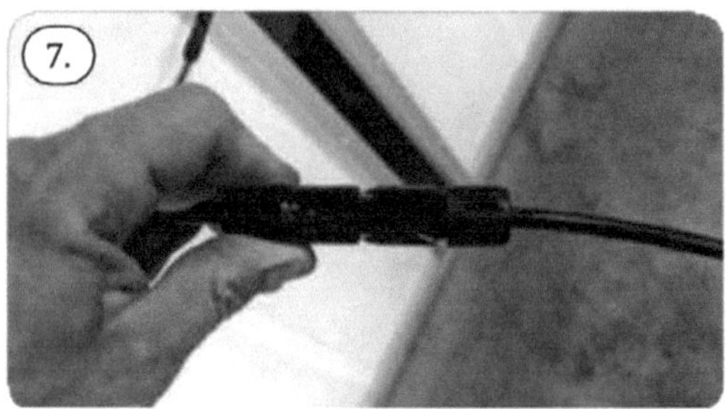

Effective Use of Ground-Based Solar Panels

The system that I have described thus far is designed with the specific intent to maximize the amount of sunlight that the panels are exposed to. The idea is that we can place panels to catch the morning Sun then re-position periodically throughout the day to keep the panels under direct sunlight. Doing this also allows you to monitor the effectiveness of your placement by keeping an eye on how much power the panels are producing.

Deploying a Portable Folding Panel

It is good to have one or two portable folding solar panels available to get some additional capacity. Deploying these panels takes minutes and requires no special builds or tools.

MATERIALS

- Folding solar panel – I used this one which is sold on Amazon for $281.49USD www.amazon.com/XINPUGUANG-Portable-Foldable-Generator-Controller/dp/B07XGKT22B/

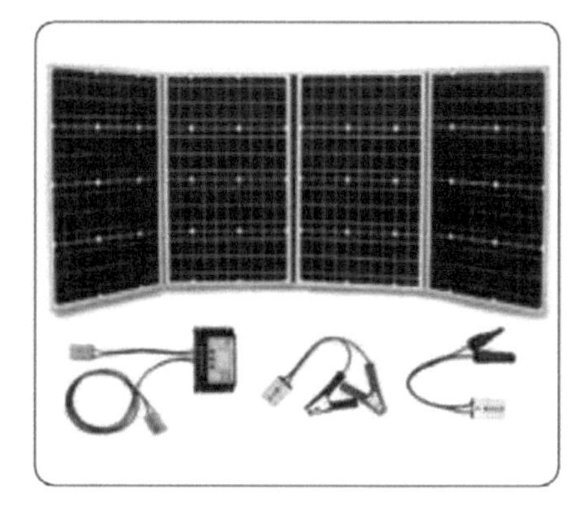

- Extension cable with MC-4 connectors – Like this one available on Amazon for $ 144.99 USD w w w . a m a z o n . c o m / W i n d y N a - tion-Gauge-Black-Extension-Connector/dp/ B01D7VBT2G/

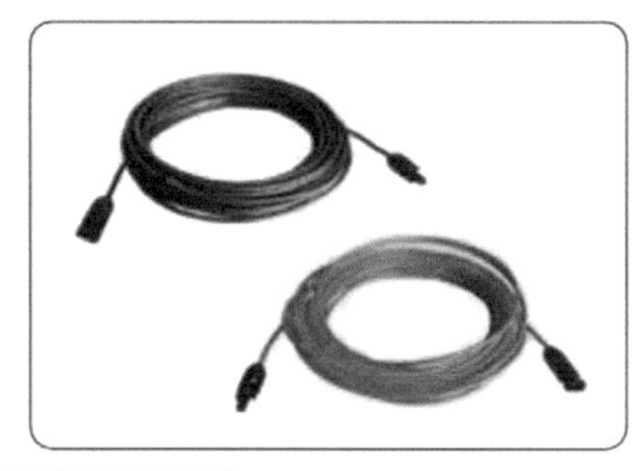

INSTRUCTIONS

1. Cut an extension cable in half the same as you would for the larger panels.
2. Mark which end is positive and which end is negative, then connect to the charge controller.
3. Connect the folding panel to the cables, always connecting the positive end first then the negative end. (You can use two uncut extension cables to connect the portable panel to your deployed main panels. This, however, may reduce the versatility of the folding panel.)
4. Unfold the panel so that it is exposed to the maximum levels of sunlight.
5. Check the charge controller to monitor the amps coming off your solar panels to tweak their placement to pull the maximum power out of the Sun's rays.

Connecting the Solar Power System to the Power Grid

In some areas, you can connect your solar power system to the power grid, which allows you to sell the excess power that you generate back to the power company. Exercising this option requires that you have permanently installed panels and that they, and the battery bank, are connected to your home's electrical panel.

Connecting a solar power system to the grid and your home's electrical system is outside the scope of an individual to DIY. There will be codes and regulations to follow, and you should only do so by using a qualified and licenced electrician and obtaining the proper permits.

Putting Your Solar System to Use

In this example, we will be powering a refrigerator, but the steps involved will be the same regardless of the appliances you are powering.

1. Check that the battery has sufficient capacity to provide power.
2. Turn on the inverter.
3. Plug an extension cord into the inverter.
4. Plug the fridge or other appliance into the extension cord.
5. Confirm that the fridge is indeed operating correctly.

How To Dry Meat And Turn It Into Powder

MEDIUM  DIFFICULTY

If you are a fan of beef jerky, and really who isn't, then you must try Jerky powder. You may have seen it before at the gas station without really knowing what it was. It is usually in little cannisters the same size as tobacco chew. And though you might have assumed it was some strange-flavored version of tobacco, don't worry, there is no tobacco in jerky powder. It's just pulverized beef jerky.

You can have it just as it is. Grab a pinch full and put it under your lip and enjoy as a burst of beef jerky goodness melts away in your mouth. It is also great as an additive to all sorts of food. Use it wherever you might throw in beef stock or bacon bits. It gives smoky richness to stews, it's fantastic mixed into mashed potatoes, and even into your macaroni cheese.

And like just about everything else, the homemade stuff is much tastier than the store-bought kind. You can tweak your recipe a thousand different ways, depending on the cut of meat you use and the sauce you marinade it in. The only consideration you must take into account when making beef jerky powder, as opposed to regular beef jerky, is that in order for you to get a powder-like consistency, the meat should be on the leaner side and the marinade should contain a minimal amount of sugar or sugar-like ingredients. This is because in order to blend up your jerky it needs to be dry. Fat, which is greasy, and sugars, which are sticky, are going to make getting your jerky dry enough difficult.

With that in mind, below is my favorite recipe for Jerky powder. If you have ever made jerky before, then you already know half of what you need to make powder. Whether you use your oven or a dehydrator is up to you, but I am a bit of a minimalist so here I am going to use a regular oven.

First, I will go over the marinade, then the oven drying method, then how to cut it up and shred it into a powder. Now when I say powder, I do not mean we are going to make jerky flour, but rather something more the consistency of sawdust. That might not sound delicious, but trust me, don't knock it untill you've tried it.

Preparing the Marinade

Generally, I am a fan of all kinds of marinades from your sticky honey soy sauce types to your more dry and peppery varieties, but again keeping in mind that this jerky is going to become powder, the marinade I am going to use here is closer to the dry and peppery spectrum.

This recipe is for about 2 pounds of beef. If you are making your jerky with more or less, adjust the recipe accordingly. All the ingredients for it can be easily found at any neighborhood grocery store.

Ingredients:

- 2 pounds lean beef cut (I use the top side, you can also use silver side, or if you want to be fancy, flank steak)
- ⅓ cup soy sauce
- 1 tsp. allspice
- 1 Tbsp. liquid smoke
- 1 tsp. chili powder
- 1 Tbsp. garlic pow- der
- 1 tsp. onion powder
- 1 tsp. black pepper

Preparing the Meat

1. First, cut your jerky into thin long strips. The thinner the better.
2. Mix all the marinade ingredients together in the zip lock bag.

3. Put the cut-up beef into the zip lock bag. I find putting the bag marinade into an ice cream container helps to stabilize everything.
4. Mix well and put into the refrigerator overnight or at least for 12 hours. I find it best to put the bag into a container in the refrigerator on the off chance that if the bag bursts open, you don't have a mess. Also, it is best to turn the bag over a couple times so that the beef inside is evenly marinaded in the sauce.

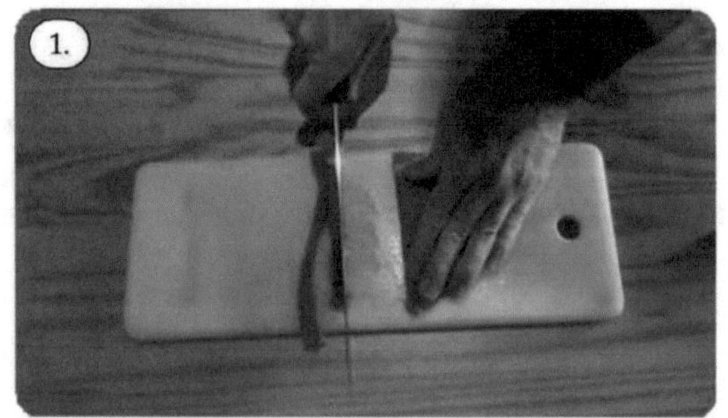

Drying Your Jerky

1. Take an oven tray and cover it with foil.
2. Place the oven rack on top of it and lay out your beef strips onto the rack. You can space them close together. The beef shrinks.
3. Set your oven to about 320 degrees and place your tray with beef into the oven.
4. Leave the oven door cracked open using a spoon.
5. Now it's time to wait. Your Jerky should be ready in about 2 hours, but it all depends on how thin you cut the meat.
6. When your jerky is dry to the touch, almost crispy, take it out and let it cool.

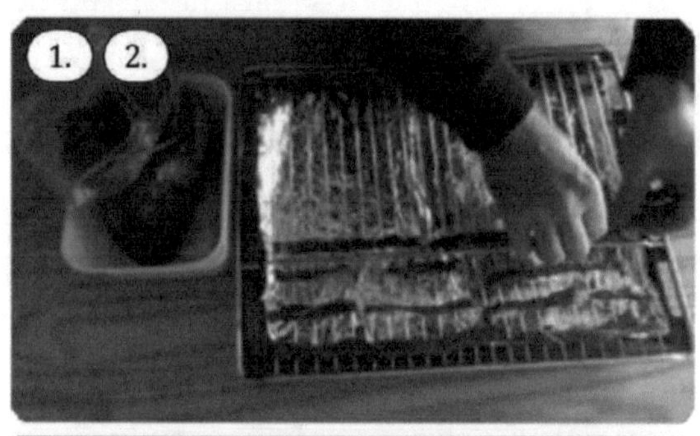

Making the Powder

1. Once your jerky is cool and dry, dice it up roughly with a knife.
2. If it feels soft while you are cutting it, you can place it back in the oven for another 20 minutes at the same 320 degrees.
3. If it cuts easily, then you can throw it straight into a food processor or blender and pulse it to your desired consistency. I like mine to still be a bit chunky, again sawdust consistency, not powdery.

If you want your Jerky to be closer to a powder consistency, maybe to use in a smooth gravy, all you need to do is blend it more. If you still aren't getting the consistency you want, try cooking your jerky longer. But be careful there is a fine line between delicious tasting jerky and burnt strips of meat. I find that the consistency this recipe produces is perfect from most applications, even gravy.

If you keep it in an airtight container, your Jerky powder should last you a good six months without refrigeration and longer with it. Enjoy!

PROJECTS ON FOOD

How To Make 2400 Calorie Emergency Ration Bars Designed To Feed You For A Full Day

EASY **DIFFICULTY** **30 MINUTES** 

Across the table from a government worker I was discussing the importance of emergency food. The discussion was centered on freeze dried meals and canned foods.

We were discussing the feasibility of both in a serious disaster situation. We were not talking about the novelty power outage where we all have fun bringing out all the emergency preparedness tools and toys.

Rather the situation where we are helping neighbors, fighting oncoming floods or trapped by the fallout of a life-threatening disaster.

It was in that moment he stressed the importance of convenience in a disaster. His example was cereal and shelf stable milk.

He described it as follows, "It takes no time at all to put together and will sustain your family members without complaint." I would be lying if I said this didn't change my point of view on disaster foods. 'Could *it be so simple*?' I thought to myself. More importantly, I began to realize how necessary this convenience could be.

Of course, there is a food that presents the ultimate in convenience and perhaps the best choice in a situation like this is the high calorie emergency ration bars.

These bars are often built in a **2400-calorie pack that is designed to feed you for a full day**. The rations are often broken into 4 squares of 600 calories each.

These rations are not only used by preppers and survivalists, but backpackers and hunters utilize them as well. This is a testament to their efficiency as a calorie provider. Of course, the elk hunter wants a delicious back strap for dinner but these rations are a nice second option.

Below I will outline the process of creating your own. If you follow the steps, you will have your own answer in a disaster scenario, or something to take on your next hike.

Materials and Ingredients Needed

As far as tools go, you will need:

- 12-inch-deep baking pan
- 1 wooden spoon
- 1 small saucepot.

As far as ingredients go, you will need:

- 3 Tablespoons Olive oil
- 2 Cups Maple syrup
- 4 Tablespoons Raw Honey
- 2 Tablespoons Peanut butter
- 1 Cup Frosted Flakes
- 3 Cups Oatmeal
- 1 Cup Protein powder
- 1 Cup Almonds
- 1 Cup Raisins.

Making Your Own Ration Bars

1. Begin by combining your honey, olive oil, and maple syrup in a sauce pot. Heat this mix over a medium heat and stir it frequently until it begins to simmer.
2. Add your two tablespoons of peanut butter to the mix in the pan. Stir the peanut butter until it melts into the syrup mix. Be careful! this syrup mix will be very hot and if it gets on your skin it's nearly napalm!
3. Take the remaining dry ingredients and add them to a large bowl or two large bowls. You don't want these bowls to be filled more than halfway, as you will be doing a lot of mixing in these bowls. If they are too full with just the dry you will have a terrible time mixing in your liquid in the next step.

4. Once you have thoroughly mixed up your dry ingredients, take the hot syrup peanut butter mixture and add it into your dry ingredients. While it's still hot, mix to coat your dry ingredients thoroughly. Make sure it's thoroughly mixed and all ingredients have a nice sheen to them.
5. Preheat your oven to 375 degrees. Dump your mix into a baking pan. This pan should be at least 2 inches deep. Be sure to press and pack this mix down tight. This will allow for tight squares to be cut from this mix.
6. Bake in the 375-degree oven for 20 minutes until the edges begin to brown.
7. Allow the mix to cool and cut into 2×2 squares. Each square will be roughly 600 calories. Packing together 4 of these squares will equal 2400 calories and be enough calories to push through a long hunting trip or life-threatening disaster situation.
8. You can even portion them in little muffin pans, if you want to get fancy. I like the little pucks to be honest.

How To Make Dandelion Bread

EASY **DIFFICULTY**

Dandelion is one of the more recognizable plants in the world. Often considered a weed to the urban gardeners, dandelion is a very nutritious plant with many medicinal benefits. Every part of the plant is edible, from the roots to the flowers.

Dandelion is a rich source of vitamins, minerals, and antioxidants. The raw greens contain high amounts of Vitamins A, C, and K while also being a source of calcium, potassium, iron, and manganese. They also have some vitamin E and small amounts of some B vitamins. Dandelion root is a source of inulin, a type of fibre that helps support healthy gut bacteria. The root is a great non-caffeinated substitute for tea or coffee.

Dandelions are generally considered safe to eat, but the risk will never be zero as with any foraged food. There is the possibility of allergic reactions in people who are allergic to plants such as ragweed, and people with sensitive skin may develop contact dermatitis.

If you choose to consume Dandelion, do so in very small amounts at first, and if you react unfavourably to it, do not consume it again. You can also do the universal edibility test to determine how you will respond to this plant.

When harvesting dandelions, only do so from areas you know have not been sprayed with pesticides or other chemicals. Your yard is probably the best area to harvest. Since we only need the flowers, you can pluck the flower head off the stem. You are going to want enough to get about a cup of petals per loaf of bread.

Dandelion bread is very easy to make and requires only six store-bought ingredients. Baking a loaf of Dandelion bread is also a good activity to get kids active in the kitchen.

Ingredients

The ingredients can be found at most grocery stores, and the Dandelion flowers can be harvested from any area free of pesticide use.

- One cup of clean, fresh Dandelions
- 2 cups of flour. Flour is $3.26USD for a 5-pound bag at Walmart

www.walmart.com/ip/King-Arthur-Flour-Unbleached-All-Purpose-Flour-5-lb-Bag/10535106

- 2 tsp of baking powder. A container of Baking powder is $1.82USD at Walmart

www.walmart.com/ip/Rumford-Double-Acting-Non-GMO-Baking-Powder-8-1-oz/252551170

- 1 egg. A dozen eggs are $4.35USD at Walmart

www.walmart.com/ip/Great-Value-Large-White-Eggs-72-Oz-36-Ct/142616435

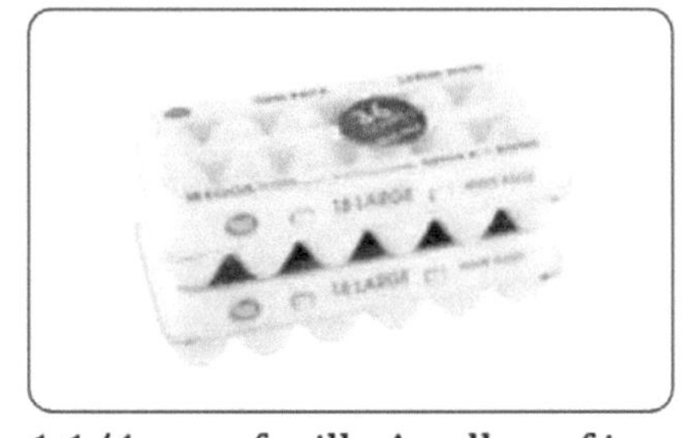

- 1 1/4 cup of milk. A gallon of is available from Walmart for $3.45USD

www.walmart.com/ip/Great-Value-2-Reduced-Fat-Milk-128-Fl-Oz/10450115

- 3 tbsp vegetable oil. Vegetable oil is $2.00USD at Walmart for a 48 fl oz container

www.walmart.com/ip/Great-Value-Vegetable-Oil-48-fl-oz/10451002

1/3 cup honey. A twelve oz jar of honey is $ 4.18 USD at Walmart

www.walmart.com/ip/Busy-Bee-U-S-A-Honey-12-oz/10292906

The Step by Step Recipe

1. Preheat oven to 400 degrees Fahrenheit and line a loaf pan with parchment paper.
2. Using a sharp knife, cut the petals off the heads of the Dandelions that you've harvested.
3. If you wish, you can keep the petals whole or chop them as fine or as coarse as you want to.
4. Mix the flour, baking powder, and Dandelions in a large bowl along with a pinch of salt.
5. In another bowl, whisk together the oil, egg, milk, and honey.
6. Pour the milk mixture over the dry mixture and stir until incorporated. Do not over-mix.
7. Pour the mixture into the loaf pan and bake for 15 minutes, then reduce the temperature to 350 degrees Fahrenheit and bake another 20 minutes until a toothpick inserted in the middle comes out clean.
8. Remove from oven and allow to cool slightly before slicing.

Dandelion bread can be stored for up to five days in an airtight container, but it can also be frozen for longer-term storage.

Taste Test

Dandelion bread has a taste that was slightly sweet and was not overpowering with Dandelion flavour. I would say that this bread would be good with some butter or honey alongside morning coffee or tea. The absence of processed sugars in this bread is very apparent, but the honey makes up for it quite nicely. I think Dandelion bread is a great way to add Dandelions to your regular diet. Making a loaf or two of Dandelion bread is a great way to introduce this plant into your diet. Baking it into a loaf of bread will help introduce the plant to picky eaters and give your family the possible health benefits of the plant, while being very de- licious.

How To Make Hardtack (Emergency Survival Bread)

EASY  DIFFICULTY

There's an old adage that says, "Man cannot live by bread alone." (I Googled it, and it turns out it's from a Bible verse.) So, when I stop to think about what foods I would want to have available in an emergency, of course, bread rose to the top of the list.

But if you know anything about bread—baking it, eating it, storing it - then you know it's prone to going bad quite quickly. And that's due to several factors, starting with yeast. But butter, oil, and milk can also be the culprits that cause your favorite loaf or rolls to mold over and rot. Any bit of moisture—the moisture that gives it loft, makes it airy, and that makes it a perfect vessel for sandwiches, cheese, and so much more—is at the root of why you cannot preserve bread for the long term. That and a lack of acid.

I figured, though, that there has to be some way to add bread to my emergency pantry.

Of course, you can freeze bread. And if you're committed to fresh bread for the long term, that's likely your best option. But if your electricity is out, then you're out of luck. That bread will eventually mold and go bad.

Then I thought, maybe I can make bread in a can. And I'll just say, while I found a couple of recipes online that explained how to do this (apparently giving gifts of canned breads, especially fruit breads, as holiday gifts was a thing a couple of years ago), I also found literally dozens of articles that said: **"No. Absolutely positively do not can bread or eat canned bread."** (Boston brown bread seems to be an exception, most likely because it's made in giant manufacturing facilities with who knows what preservatives. It is tasty, though, so it's a viable option).

The bottom line is that canned breads are not safe to eat. Trapping moisture and oxygen in a glass jar, it turns out, delivers an environment that is perfect for breeding microorganisms that can be not only harmful but that have the potential to kill you. Botulism is a prime candidate for scary things that can grow in your canned bread.

According to Richard Andrew, a food safety and nutrition educator with Penn State Extension, "There are no reliable or safe recipes for baking and sealing breads and cakes in canning jars, and storing them at room temperature for extended lengths of time." I don't know about you, but for me, botulism isn't worth the risk.

I did, however, find one viable option, and it's been around for centuries. Have you ever heard of hardtack? Hardtack is also known as survival bread, and with this food item, the term bread is used ... loosely.

Hardtack was used on ships as they crossed the Atlantic or navigated to and from exotic lands hundreds of years ago. It was also popular with pioneers and settlers as it can, when properly stored, last for years.

It's also ridiculously easy and inexpensive to make. Just be sure to temper your expectations, because hardtack is probably like no other bread you've ever eaten. It's not soft, fluffy, or flakey. It bears no resemblance even to the driest, most crisp cracker you've ever eaten.

Hardtack is, in a word, hard, and in another word, bland. It's so hard that to eat it, it must be soaked or submerged in some form of liquid for at least 5 to 10 minutes. Without soaking it in water, milk, soup, stew, brine, coffee, tea, or some other liquid, hardtack is virtually inedible. It was created specifically to be a long-lasting survival food that would not spoil.

In addition to being long-lasting, hardtack may also thicken the liquids to which it is added, making them more filling and satisfying, which is always a good thing in a survival situation. So let's get cooking, shall we?

What You Will Need

To make hardtack bread, you will need the following items:

- measuring cups and spoons
- mixing bowl
- fork or wooden spoon
- baking tray or cook- ie sheet
- rolling pin
- knife

- skewer or chopstick
- optional: biscuit cutter or jar lid

You will also need the following ingredients:

- 2 cups of all-purpose flour
- ¾ cup of water
- 1 ½ teaspoons salt
- optional: dried herbs or seasonings

With these ingredients in these proportions, you can expect to make 10 to 12 "biscuits" or "crackers." When I made the recipe, I ended up with about 14 total pieces of hardtack.

As you can tell from the ingredients list, it's very easy to double or even triple this recipe, if you want to make a lot all at once.

The Step by Step Recipe

STEP 1

Preheat the oven to 375^0F. Combine the flour and salt in a mixing bowl. You want to make sure the salt is well dispersed through the flour before the water is added.

STEP 2

Add the water.

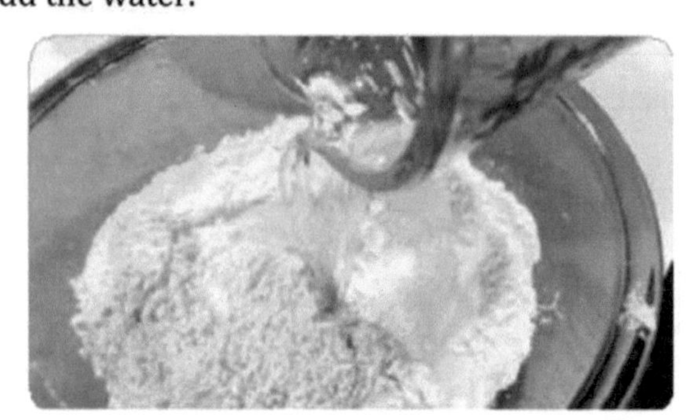

STEP 3

Mix thoroughly with a fork or wooden spoon.

The dough should be well incorporated and relatively dry but not sticky.

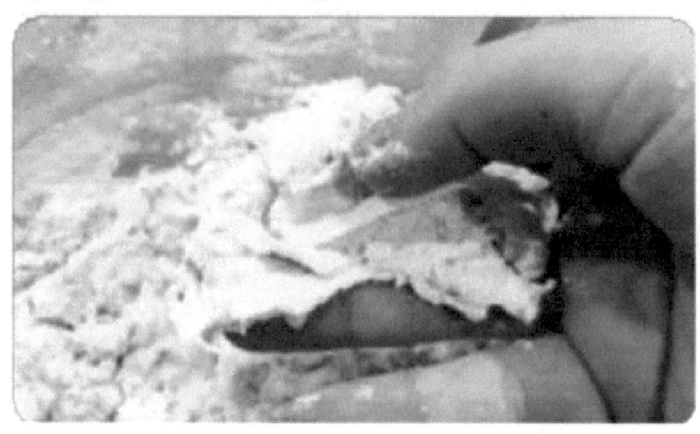

If your dough is too sticky to work with, add additional flour, one tablespoon at a time. (I did not have to add any extra flour.)

STEP 4

Turn the dough out and work it into a ball with your hands.

(My dough was crumbly, so I had to work loose flour from the bottom of the bowl into the dough ball.)

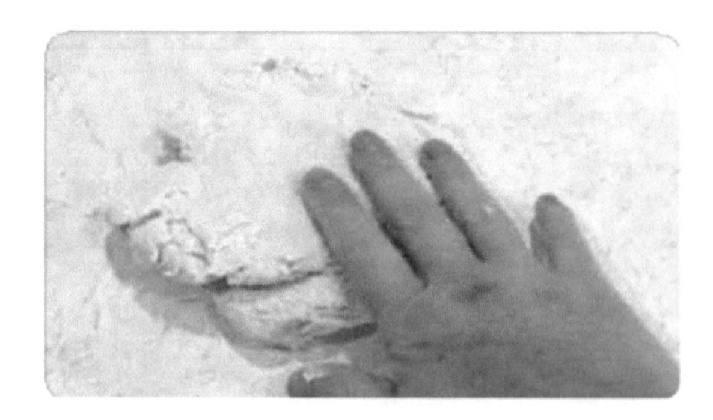

If necessary, you can add a little flour to your work surface to prevent the dough from sticking.

STEP 5

Roll the dough out with a rolling pin to a thickness of about ½ inch.

STEP 6

Cut the dough into 3-inch squares with a knife or into rounds with a biscuit cutter. (I used a canning jar lid to make rounds because my biscuit cutter is now a tool for use only in my art studio.)

STEP 7

Poke several holes into each piece of hardtack using a fork, skewer, or chopstick.

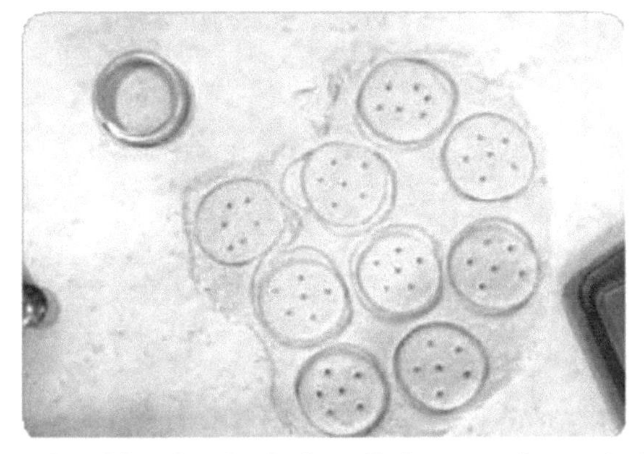

You should poke the holes all the way through the dough. This step prevents the hardtack from puffing up when it bakes.

STEP 8

This next part isn't required, but I was feeling fancy when I was testing this recipe, and if you want to add something to your hardtack to give it a little bit of flavor, feel free to do so.

Remember, you do not want to introduce any moisture to your recipe. With that in mind, I sprinkled the tops of my "biscuits" with a mix of dried Italian herbs (from last summer's herb garden).

STEP 9

Rollout any remaining dough, and again, cut out squares or rounds. This time I cut the dough into squares, poked holes into each with a skewer, and added a dried "Everything Bagel" seasoning for a little (very little) excitement.

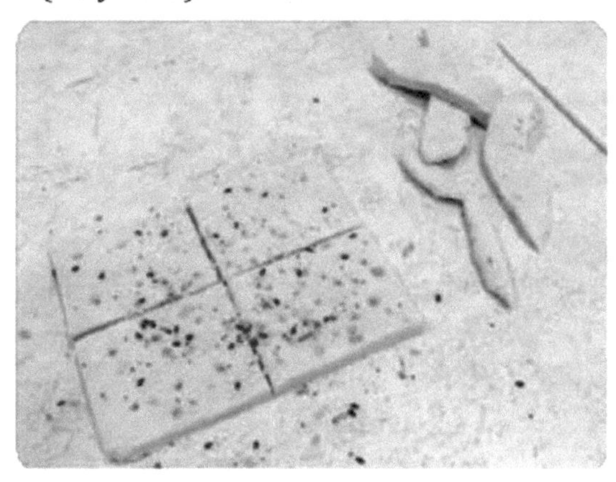

STEP 10

Place the remaining pieces of hardtack dough onto your baking sheet.

STEP 11

Place your baking sheet in a preheated 375^0 oven and bake for 30 minutes.

STEP 12

After 30 minutes, remove the hardtack from the oven (both the oven and the hardtack will be extremely hot. Use oven mitts to keep from burning yourself. Using a spatula, flip all the hardtack and return the pan to the oven for another 30 minutes.

STEP 13

Remove your finished hardtack from the oven! Here are a few photos of the finished bread:

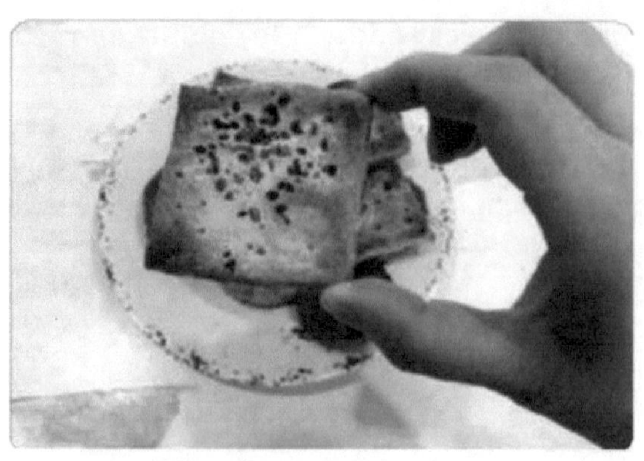

STEP 14

To ensure my hardtack will last for as long as I need it to (as well as to protect it from rodents), I'll vacuum seal it and store it in my pantry.
Don't forget to soak your hardtack before eating it!

How To Preserve Eggs

EASY **DIFFICULTY**

Eggs, one of the most versatile and commonly eaten foods in the world, are a fantastic source of vitamins, minerals, and animal protein. And they come in their own packaging which makes them great for long term storage and off the grid living. If stored in a cool place, fresh eggs that have not been cleaned, bleached, or otherwise processed can last on their own for a good month. That is because eggs fresh from the hen come naturally coated in what is called a "bloom." This bloom seals the eggshell, which despite appearances is actually porous, preventing bacteria from infecting the egg. Refrigeration of course also retards bacterial growth.

There are many other ways to store eggs for at least a year without refrigeration. Here I am going to discuss three of them and show you step by step how they are done. All of them work on the principal that with a little help the eggshell is perfectly capable of keeping its precious contents safe for human consumption. The goal is to strengthen the eggshell's natural bloom, or to add another layer of protection to keep those nasty bacteria at bay.

Preserving Eggs with Mineral Oil

You can find Mineral Oil online or in physical stores. Here are two suggestions on where to get it from:

- Amazon
www.amazon.com/Mineral-Earthborn-Elements-Cutting-Utensils/dp/B07H466M71/ref=sr_1_13?dchild=1&keywords=mineral+oil&qid=1620385761&sr=8-13

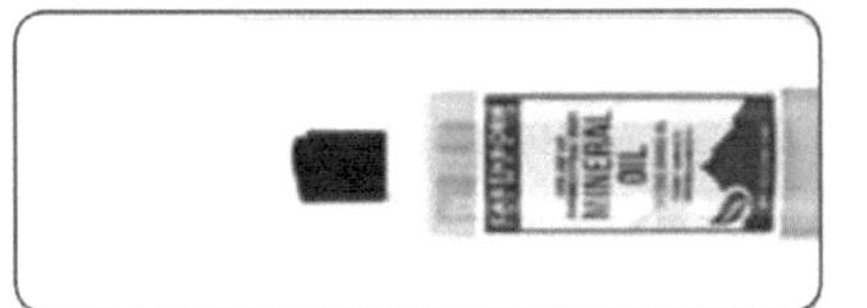

- Walgreens
www.walgreens.com/store/c/walgreens-mineral-oil/ID=prod6154213-product

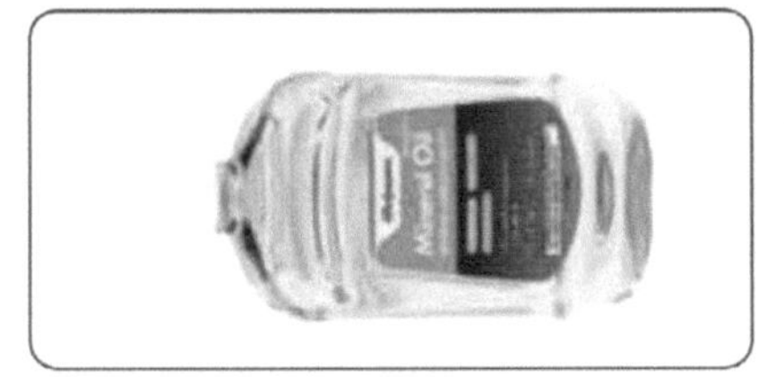

The first and most straightforward way to do this is with food grade mineral oil. Mineral oil, as opposed to vegetable or animal oil, is a by-product of the refinement of petroleum. It is used for a variety of purposes from cosmetics, medicinally (it will clear you out!), and even as a wood preservative. Accordingly, it is easy to find and quite cheap. Any pharmacy will have it and of course you can find it easily enough online. Depending on how much you buy, it will set you back around $7 dollars. Just make sure it is food or medicinal grade. As an added bonus, a little of it goes a long way. Just a tablespoon of the stuff is enough to coat about a dozen eggs. All you need to do is take a paper towel, or any piece of clean cloth, dip it in the oil, and rub the outside of each egg thoroughly. These oiled eggs, if kept in a coolish place will easily last 9 months. In fact, any oil, even lard or butter, can be used in the same manner with a similar effect. The benefit of mineral oil is that because of its purity it won't go rancid like lard or butter eventually will.

How to Do It:

First pour your mineral oil into a small bowl.

Grab your eggs and paper towel/cloth.

And start rubbing!

Place the coated eggs back in their crate and into a cool place.

"Waterglassing" Your Eggs with Pickling Lime

You can find Pickling Lime online or in physical stores. Here are two suggestions on where to get it from:

- Amazon
www.amazon.com/Mrs-Wages-Pickling-1-Pound-Resealable/dp/B0084LZU1Q

- Walmart
www.walmart.com/ip/Pickling-Lime-16-oz-Zin-524974/891358416

Waterglassing your eggs involves submerging them in a bath of water that has been mixed either with sodium silicate or calcium hydroxide, more commonly known as "pickling" or "slacked" lime. Just as with the mineral oil, the idea is that this water bath will bolster the eggshell's own bacteria fighting capabilities. The term "waterglass" comes from the fact that silicate is also one of the building blocks of glass. Lime, however, is easier to find and cheaper, so I will be demonstrating with it. Pickling lime can be found online or even usually at any nearby big box grocery store for around $10 dollars, depending on how much you buy.

The ratio of pickling lime to water is 1 ounce lime to 1 quart water. If you use tap water, make sure to boil it as it probably has fluoride in it, which will inhibit the preservative properties of the lime. Take any pot/bucket that is big enough to hold all your eggs; I'm using a roughly one gallon one for 30 eggs which is plenty. Pour in your water and lime and stir until it is all dissolved. Then simply place your eggs in the solution arranging them so they are fully submerged. And that is it, put them in a cool place and your eggs will easily last you a year or more.

How to Do it:

Get a pot, eggs, water, and your lime.

Pour the lime into the bucket and stir until it is dissolved.

Place your eggs into the water solution, making sure that they are all submerged.

Place the bucket of eggs in a cool place.

Preserving Eggs with Isinglass

You can find Pickling Lime online or in physical stores. Here are two suggestions on where to get it from:

◆ Amazon www.amazon.com/Mrs-Wages-Pickling-1-Pound-Resealable/dp/B0084LZU1Q

◆ Sound Homebrew Supply www.soundhomebrew.com/isinglass-45-ml/

If you want to stay away from any mineral-based preservatives, then there is this final option. Isinglass is a powder that is derived from the swim bladders of fish. It is commonly used in the wine and beer industries as a clarifier because it binds itself to suspended particles and then sinks to the bottom of whatever liquid it is put into. During WWII, however, the British found that isinglass could also be used as an egg preservative. The concept is the same as that of water glassing, but instead of lime we use isinglass, which forms into a gelatin-like substance when mixed in the right proportion with water. This gelatin provides an extra layer of protection by blocking up the tiny holes in an egg's shell.

Isinglass is fairly easy to find online or at your local homebrew supplier. And as with the pickling lime, a little bit goes a long way. The ratio of isinglass to water will vary depending on the type of isinglass you get. Sometimes it comes as a clear "paper" as seen below and sometimes as a powder. Generally, the ratio is something like 1 part isinglass to 10 parts water. The goal is for the isinglass to dissolve into the water completely so that upon cooling it solidifies into a gelatin.

How to Do it:

First, heat up enough water to fully submerge the eggs you want to preserve and pour in your isinglass.

Stir until it starts to thicken.

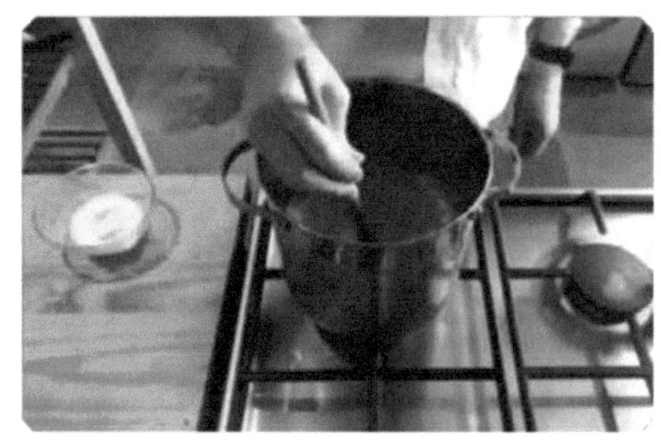

Arrange your eggs in a clean container.

Pour in your isinglass water until the eggs are fully submerged.

Wait for it to cool to the point that the gelatin has set.

That is all there is too it!

Eggs preserved in all three of these ways should also last you at least a year.

Meal In A Bag: Hamburger Gravy And Mashed Potatoes

EASY  **DIFFICULTY**

Whether you have been on a military tour or just on a tour hiking around your nearby national park for a few days, you have probably had an MRE bag meal. Now, certainly some are better than others. Many of the ones that you buy from outdoors outfitters are pretty fancy and promise a lot. But if you are like me, most of them have been disappointing.

When it comes to MREs, I have found that the ones that keep it simple tend to come out the best; so with that in mind here is a recipe that I have made for a homemade Meal in a Bag that I think comes out very tasty: good old hamburger gravy with instant mashed potatoes. While I usually try to steer clear from overly processed powders, both for health reasons and because they never taste as good as the real thing, this recipe is going to use a fair bit of pre-made ingredients. Most of them can be found easily at your nearby grocery store.

This recipe has two parts: the gravy and the mash. You will be storing the ingredients for each part separately because they don't store well together. While long term storage is usually accomplished with canning, using glass jars, here I am going to stay true to the MRE-style of using mylar bags. Mylar bags are made of food-grade plastic and aluminum foil and once sealed can keep out bad bacteria and lock in freshness. Usually when you buy them, they also come with oxygen absorbers that you put into the bag before sealing it up. These little guys are important. If you don't add them, then bacteria can grow inside the bag even after it is sealed.

What You Will Need

- Mylar bag (You can find them easily online www.amazon.com/6-5x4-Black-Mylar-Storage-

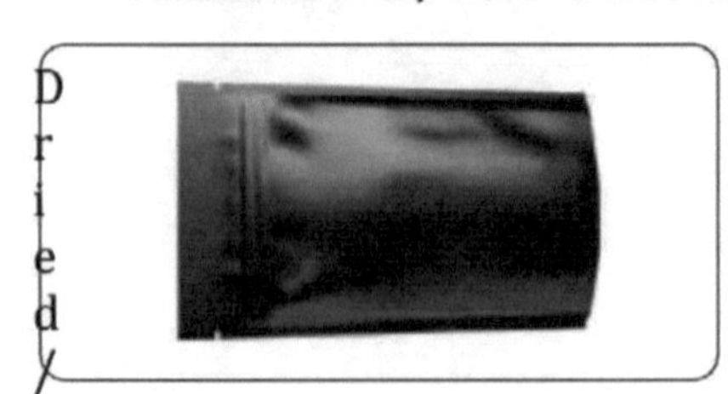

Dried/
dp/B08N5P58GC

and at your neighborhood Home Depot) www.homedepot.com/p/Harvest-Right-Mylar-Starter-Kit-with-50-Mylar-Bags-50-Oxygen-Absorbers-and-12-in-Impulse-Sealer-MYLARKIT/305561948

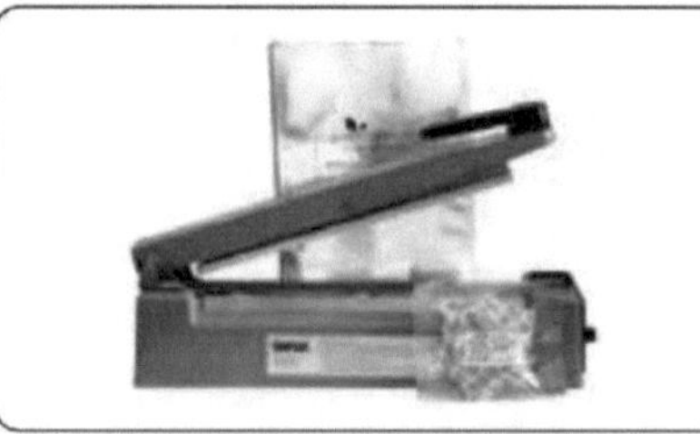

- Iron or hair straightener (This is for sealing the mylar bag.)
- ¼ cup freeze-dried hamburger meat (This one can be a little tricky to find but again you can get it online. If you don't want to order online, you could try to make your own.)
- 1 cup Instant Mash Potato powder
- 1 tsp. Beef Stock cube or powder
- 1 Tbsp. full fat milk powder
- ½ tsp. mushroom powder
- 1 Tbsp. flour
- 1 tsp. garlic flakes
- 1 tsp. onion powder
- ½ tsp. mixed herb
- ½ tsp. black pepper

The Step by Step Recipe

1. Simply place your freeze-dried ground beef into the mylar bag.
2. Mix all the other ingredients together and put in a separate plastic bag.
3. Close that bag up carefully to get as much air out of it as you can and place it into the mylar bag.
4. Seal up the mylar bag.

It really is that simple.

Now, when it comes to cooking time you can either use the bag itself as the container to cook everything in, or as I prefer, use a pot over the stove or campfire. All you need to add is 1-1 ½ cups hot water.

1. Remove your plastic bag of mash potato mix and rehydrate your ground beef either in the bag or in pot. Close/cover it up and wait 10 minutes for the ground beef to rehydrate.
2. Then pour your potato mix and stir until your mash fluffs up.
3. Enjoy!

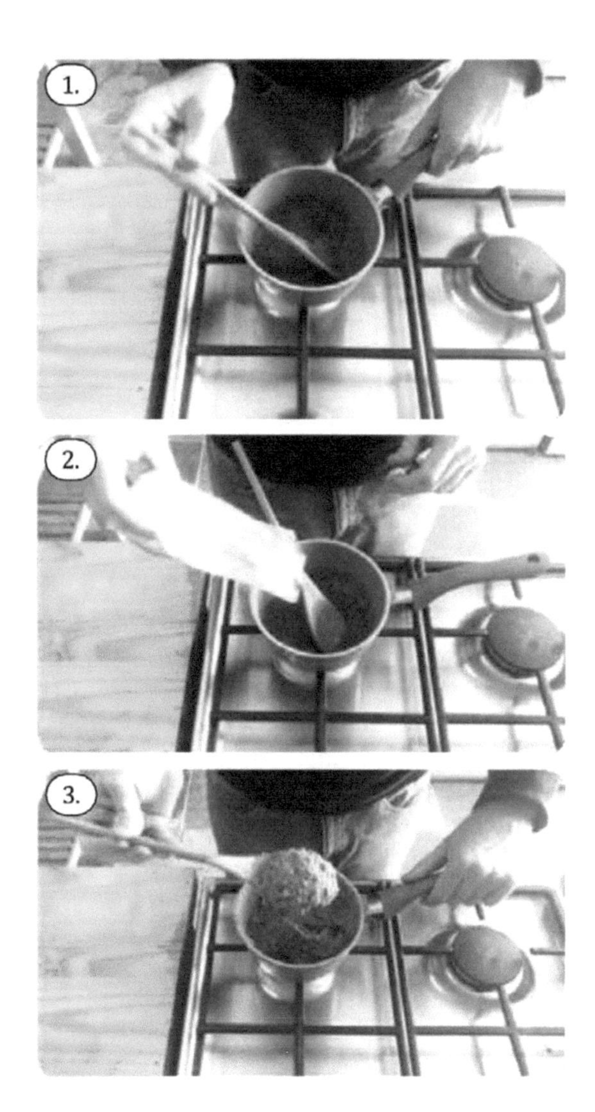

Automatic Traps For Animals

MATERIAL COST  **45.00** **MEDIUM**  **DIFFICULTY** **30 MINUTES** 

Hunting is hard. Trapping is not much easier but can work for you while you are doing other things. There is also the advantage of placing a large volume of traps, which increases the odds of success. Trapping has a steep learning curve, but as long as you pay attention and learn from failure, setting traps can be a very effective means of procuring protein.

One can easily construct many traps either in the wild or at home with readily available materials. The following are traps that I always keep in my arsenal of protein procurement if I need to use them.

Most jurisdictions have licencing requirements for the trapping of animals. In most areas, using these techniques outside of an actual survival situation will be considered illegal. Always read and understand all rules, regulations, licencing, and laws you are subject to.

DIY Wire Snare

The snare is one of the most basic trapping methods and can be very useful when constructed and deployed correctly. While it is possible to use brass snare wire or picture wire, a snare built with a more sturdy material will hopefully prevent animals from destroying the snare and escaping.

I have found that building these snares usually takes less than ten minutes per snare, and if set up in an assembly line, the time to make these snares will be much less.

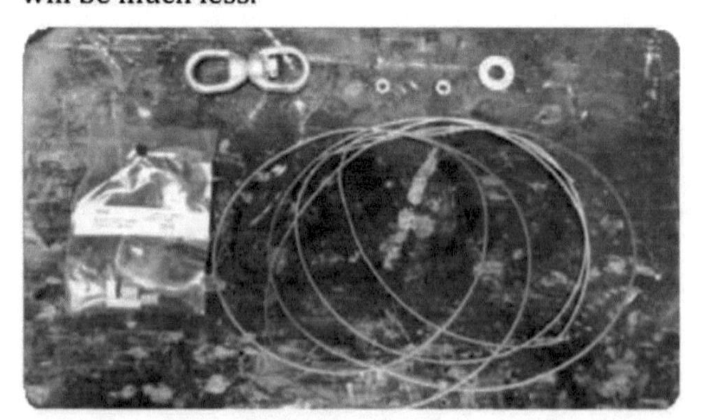

Materials

- 1/16" Aircraft cable – You can purchase this at Home Depot for $12.97USD for 50 feet
www.homedepot.com/p/Everbilt-1-16-in-x-50-ft-Galvanized-Steel-Uncoated-Wire-Rope-811072/300018981

- 1/16" Aluminum Ferrules – Available at Home Depot for $3.46USD for a pack of ten
www.homedepot.com/p/Everbilt-1-16-in-Aluminum-Ferrules-10-Pack-42574/205887624

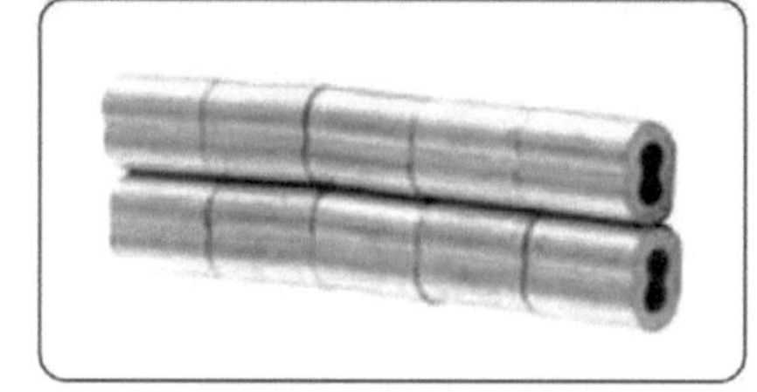

- ½" Flat Washers – Available at Home Depot for $5.94USD per pack of 25
www.homedepot.com/p/1-2-in-Zinc-Plated-Flat-Washer-25-Pack-802334/204276390

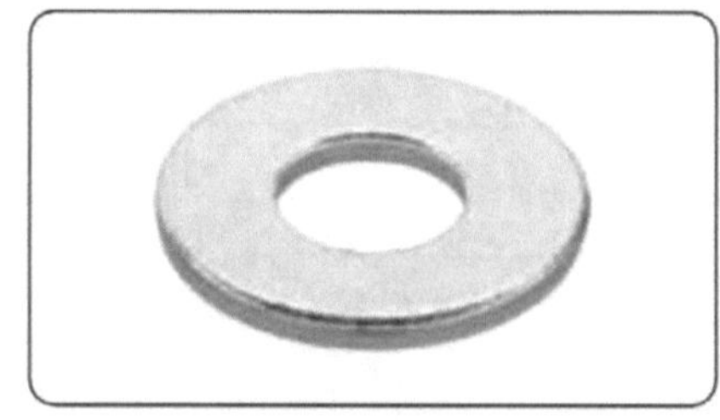

- 10-24 Nuts - Available at Home Depot for $ 6.96 USD for a pack of 25
www.homedepot.com/p/Hillman-Stainless-Machine-Screw-Hex-Nut-10-24-958/204794767

- Swivel – Available at Home Depot for $ 5.98 USD each
www.homedepot.com/p/Everbilt-0-25-in-Galvanized-Eye-and-Eye-Swivel-44094/205874115

Tools

- Drill and drill bits
- Bench Vise
- Hammer
- Good quality side cutters

Constructing the Snare

Constructing these snares is a straight forward process, and each snare will take around ten minutes to construct.

1. Start by cutting a 36-inch section of cable for each snare you want to create. From a 50 foot roll of wire, you can get 16 individual snares.
2. On one end of the wire, slip one of the 10-24 nuts and compress it onto the wire by either squeezing in a vice or smashing with a hammer.
3. Secure the washer in the vise.
4. Drill two holes in the washer directly across from each other large enough for the wire to slide through easily. I used a 3/32" drill bit.
5. Using the bench vise bend the washer in half at a 90-degree angle.
6. Run the other end of the wire through one hole of the washer, as shown in the photo above.
7. Run the wire through the other hole as shown.
8. Slip a ferrule onto the open end of the wire.
9. Slide the swivel onto the end of the wire.
10. Create a loop that secures the swivel to the end of the snare.
11. Crimp the ferrule using a crimping tool or side cutters.

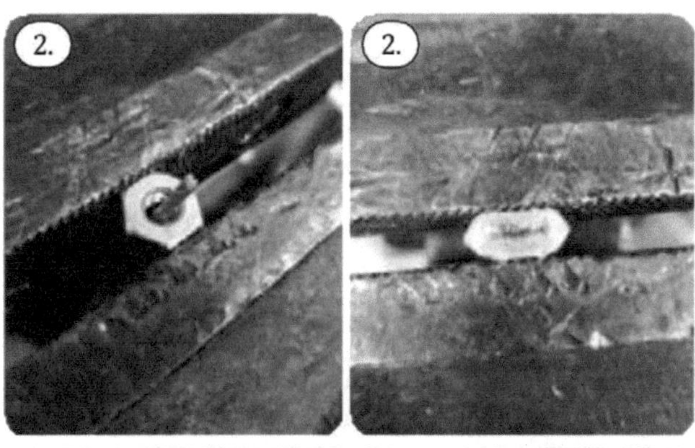

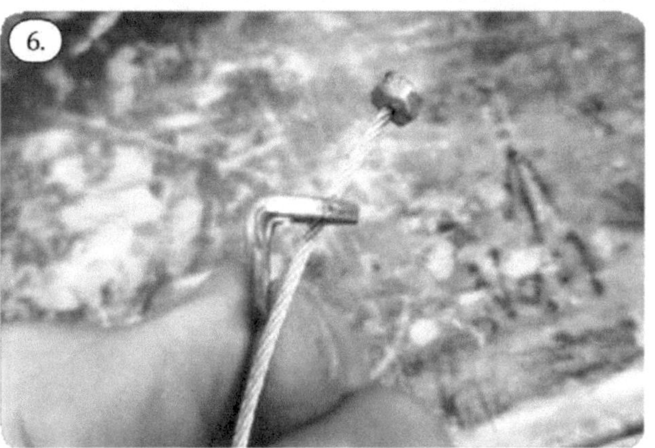

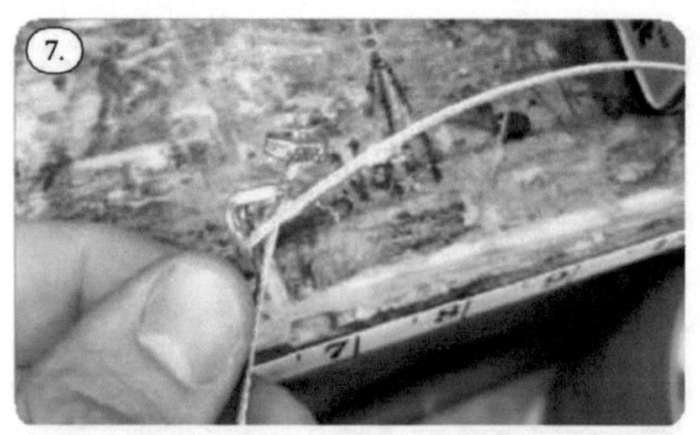

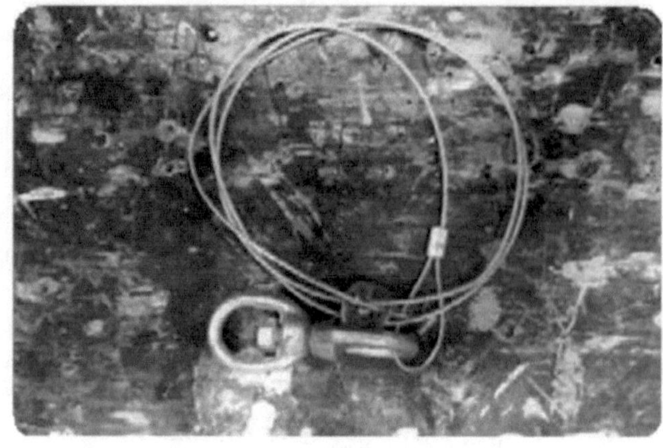

Placing the Snare

1. Find an area where your target species of animals are likely to move through. Anchor the snare to a stake or tree by securing the swivel with bailing wire.
2. Place the snare, propping it up with sticks to keep the loop open.

Twitch Up Snare

This trap uses a trigger system that is adaptable to other methods of trapping and killing game. The advantages of using a twitch-up snare are that having the snare lifted after the trap is trig- gered eliminates the need for a locking device and also elevates your dinner in the air, hopefully out of reach of any scavengers.

Tools

- Saw
- Knife

Materials

The good news is that you should be able to source everything you need from nature, but it will be far easier and quicker to use store-bought cordage like paracord.

- Paracord – Paracord is available on Amazon.com for $10.98USD

www.amazon.com/BENGKU-Survival-Mil-SPEC-Parachute-MIl-C-5040-H/dp/B07226B3FJ/

Constructing the Trap

A twitch-up snare can either have a sapling or branch as the trap's engine or a suspended weight that drops when the trap is triggered. Either method will work, so choose the one that suits the area that you are trapping.

Construction of a Twitch Up Snare

1. Locate an area that has evidence of animal movement through it.
2. Cut two branches that are about 1 inch in diameter at between 12 and 18 inches long. Sharpen one end of these branches to a point.
3. Cut three slightly thinner branches (between ½ inch and ¾ inch in diameter) at around 10 inches long and another at 4 or 5 inches long.
4. Lash one of the 10-inch braches to both the braches that you have sharpened as a crossbar.
5. Drive the sharpened branches into the ground parallel to the expected travel direction.
6. Take the 4-inch branch and carve a groove into the center of it large enough to accommodate the paracord.
7. Cut a length of paracord long enough to run from the trigger area to the sapling or counterweight you are using for this trap.
8. Tie a loop at the end of the paracord. Run the other end of the cord through this loop, creating a noose.
9. Attach the 4-inch branch to the paracord using the groove that you've carved. I like to use a clove hitch for this.
10. Tie the tag end of the paracord to either a sapling or counterweight.
11. Assemble the trigger mechanism by placing the 4-inch branch behind the crossbar. Then place the other 10-inch branch on the opposite side that the crossbar is mounted.
12. Use the four-inch branch to hold the lower 10-inch branch in place, as shown in the photo above. As long as there is tension on the line pulling up, the branches will hold themselves in place. Lay a stick on the bottom cross branch and position the snare over top of it.
13. At this point, when the lower branch has pressure applied down onto it, the branch slips down, allowing the noose and paracord to spring upwards.
14. Experiment with different setups using this trigger system to find one that works the best for you and the area you live in.

Placement of a Twitch Up Snare

These traps work well along travel corridors. Find a spot that you see evidence of animals moving through and place the trap along their route. If you bait the trap so that the noose drapes over the trigger stick, the loop will tighten around the animal's neck when the animal tries to take the bait. Alternatively, the noose will tighten around the animal's foot, hopefully elevating them into the air. Trapping is a far more efficient means of protein procurement than hunting, but it requires some knowledge and skill. Trapping is not as simple as setting snares; it requires knowledge of the local animals and their behaviour so that you can determine the most likely locations for trap sets. Take the time to learn, and get out in the bush to observe animals, do some tracking, and become familiar with their habitat because the time you will need to depend on your trapping skills to feed your family is too late to start learning.

Automatic Traps For Fish

MATERIAL COST  **31.00** **MEDIUM**  **DIFFICULTY** **1 HOUR** 

Human beings have been fishing for at least 40,000 years, and for a good reason. Fish are a fantastic protein and fatty acid source while generally requiring a low-calorie expenditure to gather. In the twenty-first century, most of us head out onto the water with a rod in hand. Still, those of us who do fish recreationally understand that the odds are not in our favour.

When our lives depend on what we can catch, gather, and hunt, betting on our skills with a spinning or fly rod is not a good wager. Instead, we need to stack the deck in our favour by using techniques that do the fishing for us, while we are focusing on other survival aspects.

While we all want to try out new techniques, what I will detail here may not be legal in your area. Before trying any new fishing method, read and understand the rules and regulations for your area. You alone are responsible for the actions you take while out on the water.

Gill Net

A gill net catches fish in three ways. Fish either become wedged in the net, their gills become caught in the net's lines, or the net tangles around the fish. This type of net is called a gill net because when a fish makes it partially through the net's openings, when they try and back out, the net's lines become caught in the fish's gills, trapping them.

A gill net can also be tailored to the size of the fish that you are looking to catch. By adjusting the openings' size, you control the size of fish that will become caught in the net. Larger openings mean that smaller fish will swim through while larger fish will become caught.

Making a Gill Net

Constructing your gill net can seem overly complicated and time-consuming. Still, it is not tricky in reality, and once you get a good rhythm going, the process does not take as much time as one would think. The 30" by 30" net I made, as an example, took about one hour for me to construct.

Materials

Constructing a gill net requires a few simple and easily obtained materials:

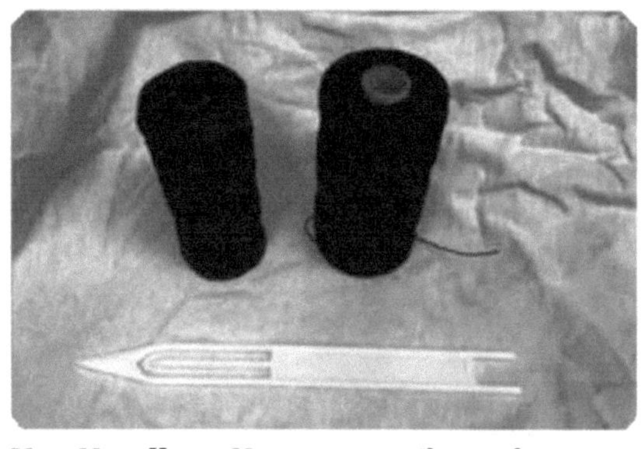

- **Net Needle** – You can easily make your net needle, but I chose to purchase some off Amazon.com for $ 7.99 USD.

www.amazon.com/Plastic-Fishing-Equipment-Netting-Shuttles-Size/dp/B0197MUEB6/

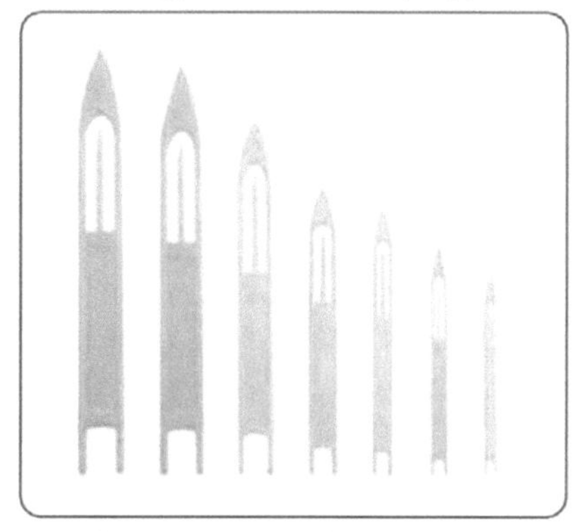

- **Cordage** – I used some tarred nylon twine (commonly known as bank line). I used the #36 bank line for the mainline, and for the actual net, I used the #18 bank line. If you wish, you can use a fishing line to make a gill net, but you'll need to make sure that it is strong enough. You can purchase some bank line rolls from Amazon.com for $22.95USD ea for either the #18 size or #36.

www.amazon.com/SGT-KNOTS-Tarred-Twine-Bank/dp/B00EVX5Y8A/

- **Mesh Gauge** – This is how you will create uniform-sized openings in your gill net. A mesh gauge can be any material that you have on hand. Wood and plastic are good options. Keep in mind that the mesh gauge width will be approximately half the mesh's finished size. For example, if you want a three-inch opening, you'll want to use a one-and-a-half-inch mesh gauge.

You can construct several gill nets of various sizes for a little more than a $50.00USD investment. You can also unravel the #36 bank line's strands to give yourself even more cordage to make more nets. Your only limitation on the size and number of gill nets you can make is the volume of cordage you have at your disposal.

Instructions for Making a Gill Net

1. String a length of your #36 bank line at about chest height. The size depends on how large you want your net to be. Leave enough excess to be able to anchor the net on either side. It is better to have too much line than not enough.

Loading the Needle

2. Load your net needle with the #18 bank line. Start by tying a clove hitch on the tip in the open centre of the net needle.
3. Wrap the bank line down and around the notch at the bottom of the needle.
4. Flip the needle over, bringing the line up to the tip, looping around it, and back down.
5. Flip the needle over again, repeating the process until the desired length of cordage is loaded on the needle.

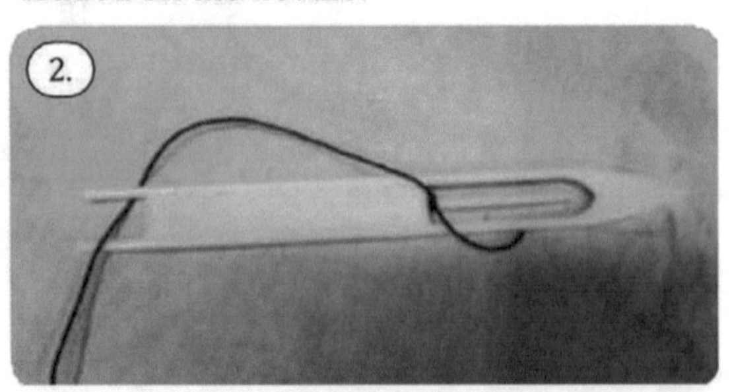

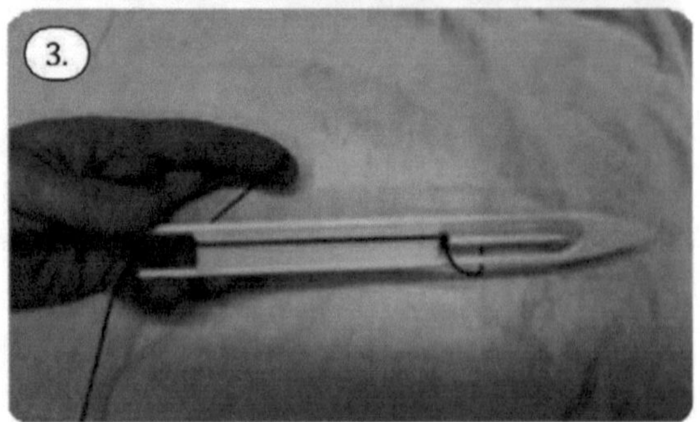

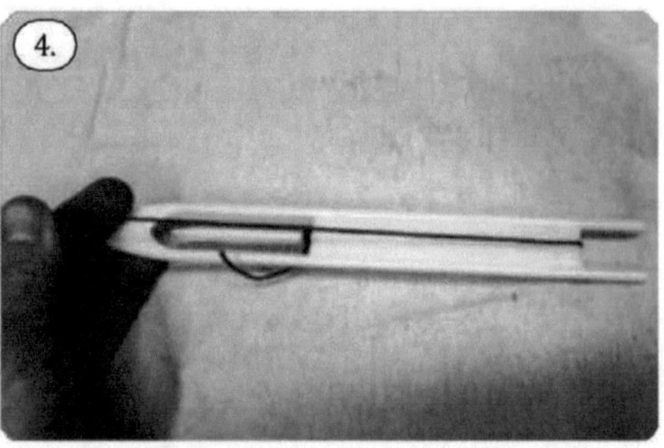

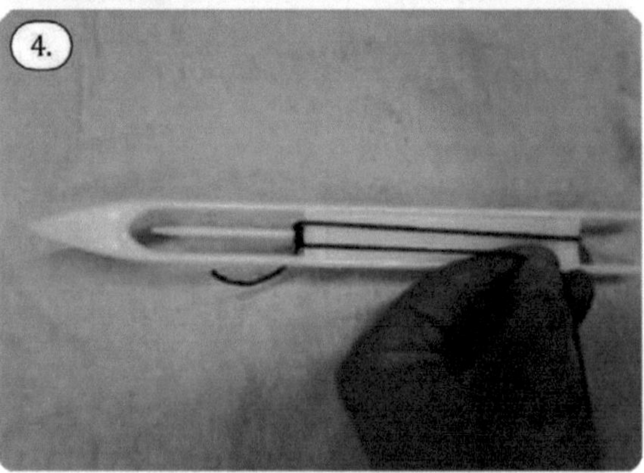

Weaving the Net

1. Tie off the loose end of the line from your net needle to the head line with a clove hitch.
2. Bring the mesh gauge up to the head line and running the thinner bank line behind, up, and across the mesh gauge.
3. Using the net needle run the line over and behind the head line. Bring the needle and line back through the loop you've created and pull snug keeping the line wrapped around the mesh gauge.
4. Repeat the last step but instead of the line being wrapped around the mesh gauge, tie it to the head line. This finishes the clove hitch that forms your first loop.
5. Repeat this procedure until you have the desired number of loops.
6. To start the next row, hold the mesh gauge below the last loop.
7. Run the net needle and line under the mesh gauge and up through the loop.
8. Pull this line down and hold it taut with a finger or thumb.
9. Run the net needle around the loop that you brought the net needle up through.
10. Bring the needle up through the loop you have made and pull tight, moving your thumb out of the way as the knot tightens.
11. Repeat this procedure for the rest of the row.
12. To make the next row repeat this procedure

back the other way.

13. Continue following these steps until the net is at the length that you want. To terminate the net, simply cut the line at the last knot. You can add a hitch or two to fully secure it or leave a tag end to fasten a weight to the net.

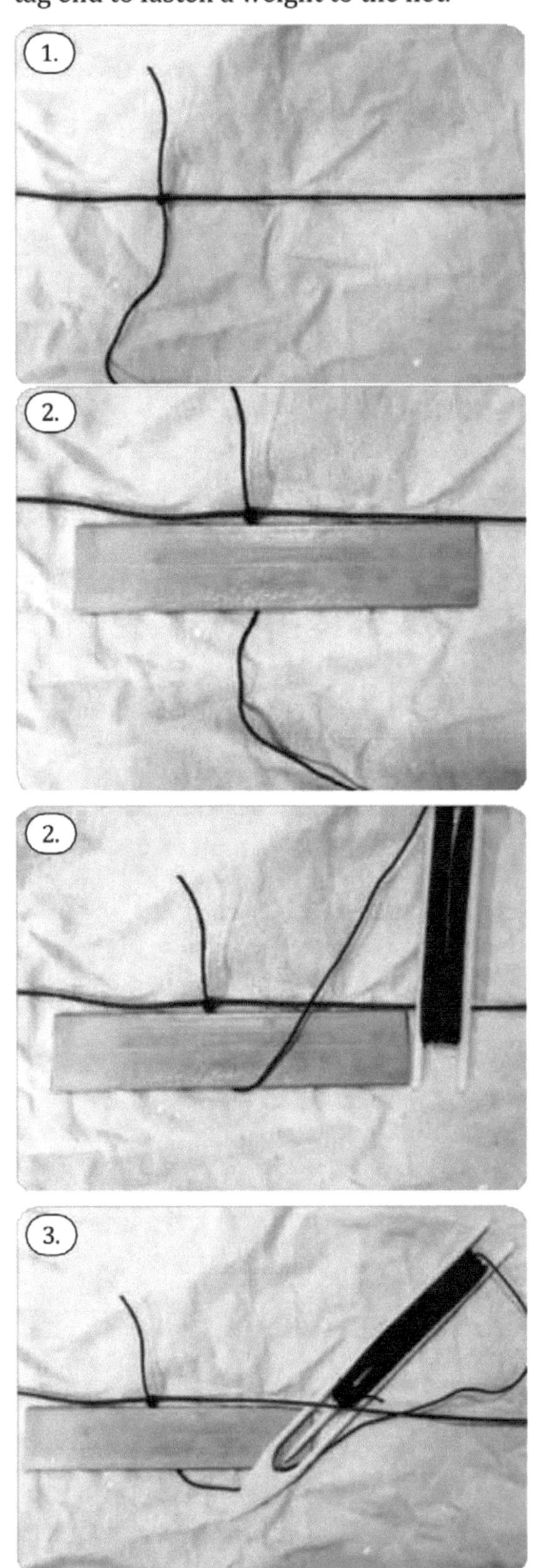

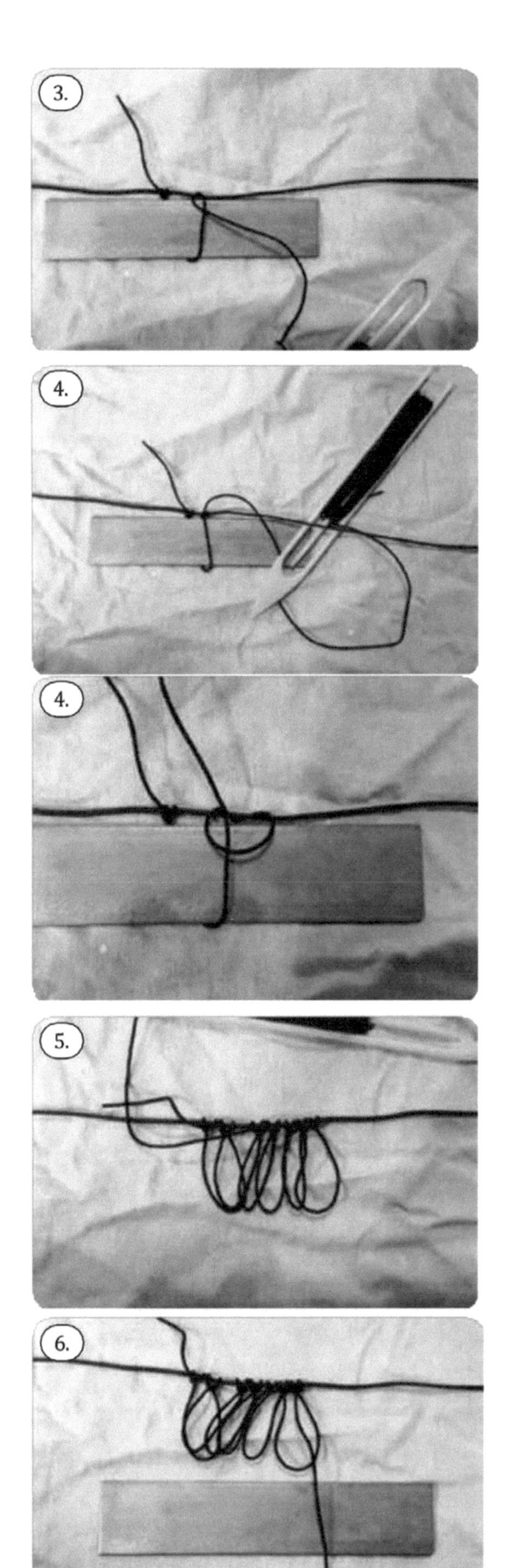

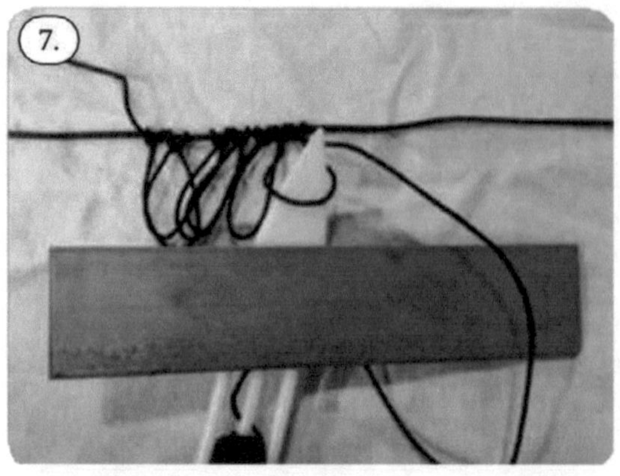

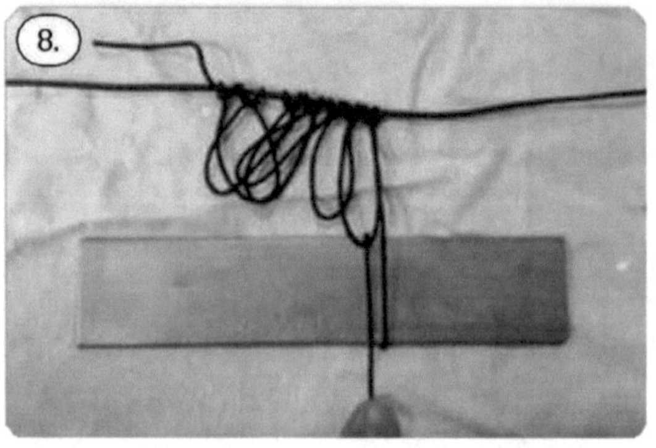

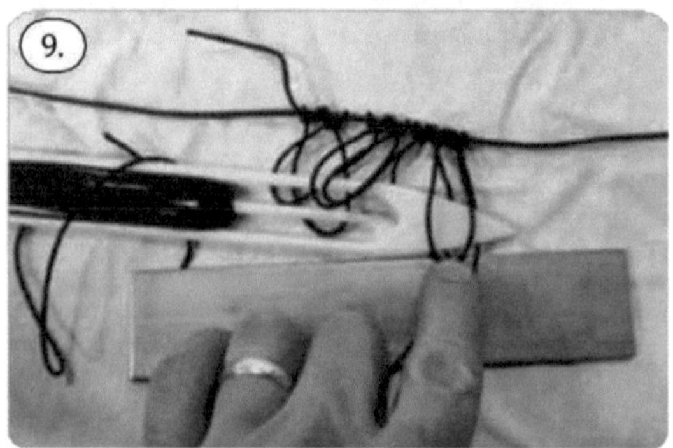

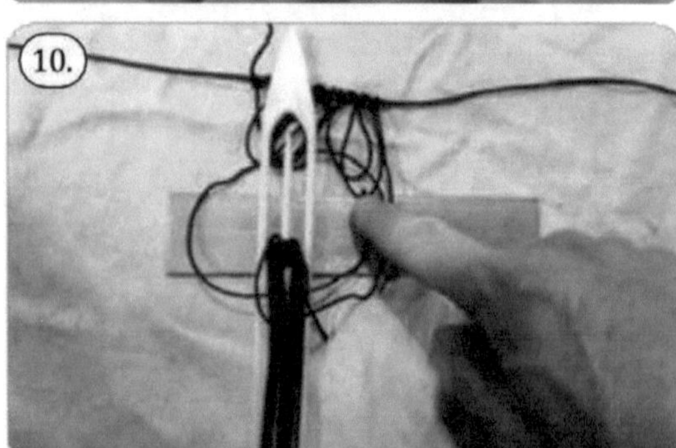

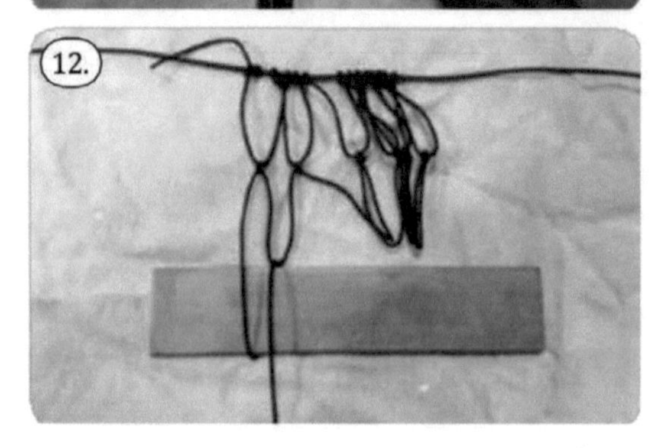

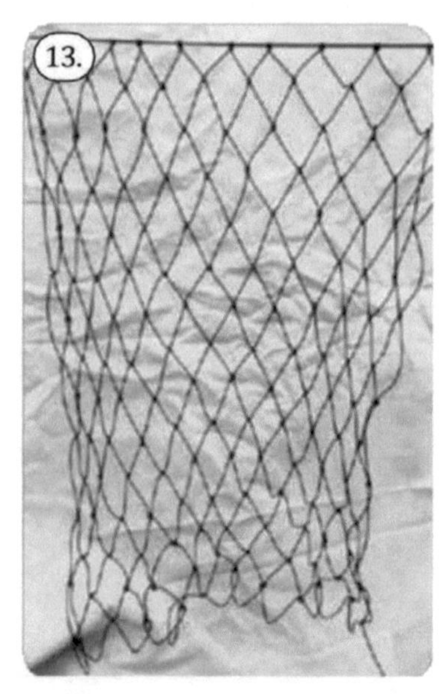

Deploying the Gill Net

Locate an area that would likely have fish swimming through it. You can secure the headline to a stick or log that you can use to extend the net over water, weighing the bottom of the net down with rocks or other weights. Suppose you have ready access to each side of the waterway. In that case, you can use the head line to suspend the net in the precise position that you want it deployed using weights to keep the net opened as wide as possible.

Trot Line

A trotline is one of the simplest passive fishing methods you can deploy aside from the gill net. A line with hooks suspended beneath it, the tro- tline does the fishing for you. The best part is that all you need is some #36 bank line and as many hooks and lengths of line that you wish to add.

Materials

- **#36 Bank line** – Amazon.com has one pound rolls for $ 22.95 USD.

www.amazon.com/SGT-KNOTS-Tarred-Twine-Bank/dp/B00EVX5Y8A/

- **Fishing Line** – This is available on Amazon.com for $ 11.38 USD for a spool of 25-pound test.

www.amazon.com/Berkley-Trilene-Casting-Monofilament-Service/dp/B0091H-DOII/?th=1

- **Fishing Hooks** – Amazon.com has collections of different sized hooks that allow you to pick the most appropriate hook for the size and species of fish you are hoping to catch for $ 12.99 USD.

www.amazon.com/Tailored-Tackle-Freshwater-Assortment-Baitholder/dp/B07C68PXG2/

Swivels – You can find these on Amazon.com in a kit that has a variety of sizes for $9.99USD.

www.amazon.com/Siasky-Connector-Stainless-Accessories-31lb-104lb/dp/B07SM4DTVL/

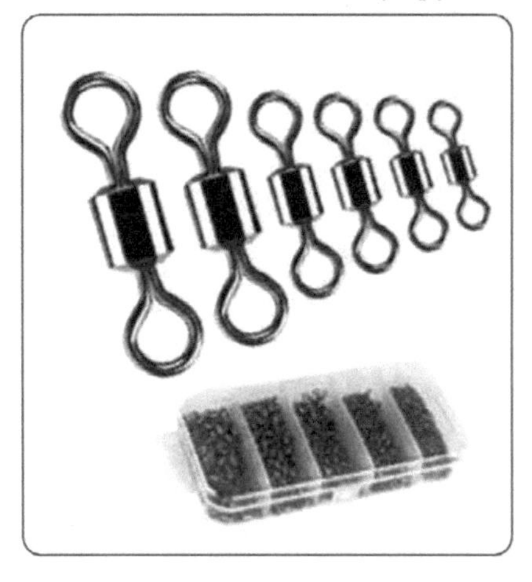

These items will give you the ability to deploy as many trotlines as a one-pound roll of bank line will allow.

Making a Trotline

Placing a line in the water with more than one hook is illegal in the area that I live, so that I won't be demonstrating the deployment of a trotline across a waterway through pictures or video. The actual deployment of a trotline depends on the waterway that you are running the line across and the fish species you are pursuing. Gaining local knowledge of fishing in your area is the best way to be successful using this technique.

1. Prepare your main line by tying swivels into it using either half hitches or alpine butterfly loops.
   Keep the swivels at least four feet apart to prevent tangles. If you are planning on running long drop lines, make the spacing greater.
2. Cut and attach hooks and bait your drop lines. You can leave a loop in one end to make attaching the drop lines to the mainline easier.
3. Attach one end of the line to an anchor point.
4. Deploy the line across the waterway by using a boat or wading, attaching the dropline to your swivels as you go.
5. Secure the other end of the mainline to an anchor point on the opposing shore. I like to use a clove hitch for this. Tie a marker on the line where your trotlines cross the shore so they do not pose a tripping hazard.

When checking and rebaiting your trotline, don't be afraid to adjust the depths of the hooks. If it is not catching fish or if your bait is left untouched, change your setup.

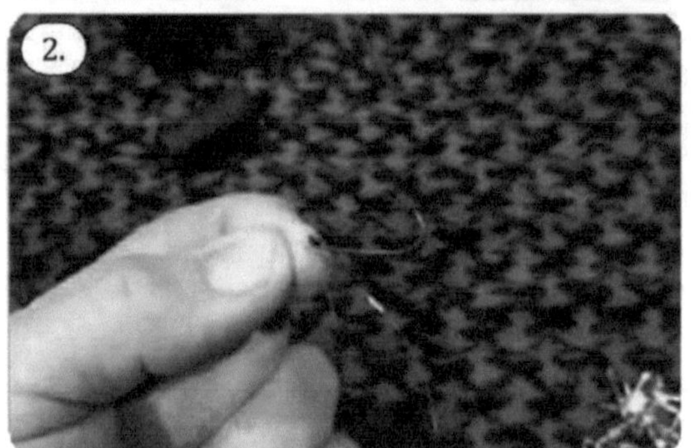

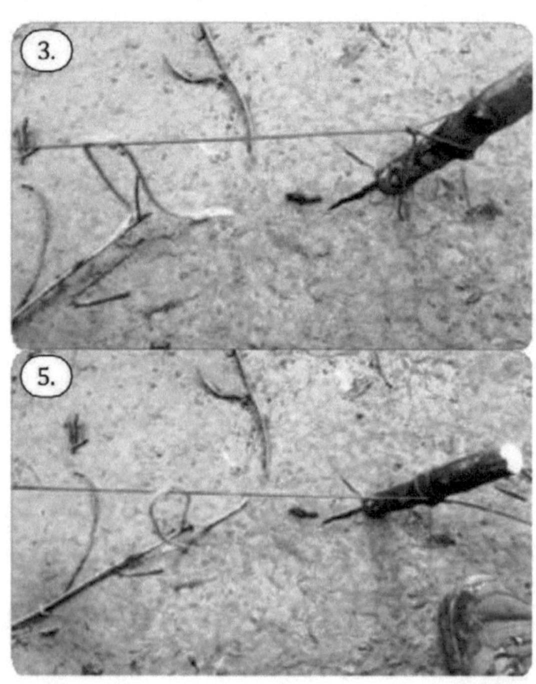

Fishing is a fantastic way to gather protein in a survival situation. Using these methods allows you to deploy several traps and cover large areas of water. This is far superior to procuring individ- ual fish using a rod and reel and allows for better allocation of your time resources.

Automatic Trap-System For Birds

MATERIAL COST  **49.00** **EASY**  **DIFFICULTY** **20 MINUTES** 

In an off-grid survival situation procuring food will be challenging. One of the most difficult aspects of obtaining protein is that there may be days where you will have good luck in hunting and trapping, but there will be more days where the wild game will be too elusive. However, an old saying goes, 'live meat never spoils,' and this trap will capture birds without harming them, providing you with live protein.

The advantage of this is that you can save the live birds for times when other protein sources are scarce and, in the process, grow them into larger, healthier birds.

How This Automatic Trap for Birds Works

This trap uses a simple mechanism to slowly lower birds into a cage, after which a counterweight resets the trap for the next bird. The cage is large enough to hold several large birds, and the top should be made from a solid material so the birds can walk around the roof of the cage with ease. The mechanism by which the birds find their way into the trap is a five-gallon bucket attached to a simple lever and counterweight.

The bucket has cutouts for the bird to step into and is suspended by a rope connected to the lever counterweight mechanism. The bucket is suspended above a hole in the roof large enough for the bucket to drop into.

The bird steps into the bucket and the bucket lowers into the cage. When the bird steps out of the bucket once in the cage, the counterweight returns the bucket to its starting position, trapping the bird in the cage.

To entice the birds into the trap, one could use birdseed, nut butter, grains, or other favourite foods of your chosen prey.

Building the Automatic Bird Trap

The construction of the automatic bird trap is a straightforward process requiring materials that you could easily purchase from a hardware store or scavenge in a grid-down scenario.

Tools Required

- Saw
- Hammer
- Sharp knife
- Drill with drivers and bits
- Tape measure and square

Materials

- 2 x 2 x 8' lumber which you can find at Home Depot for $3.48USD.

www.homedepot.com/p/2-in-x-2-in-x-8-ft-Furring-Strip-Board-Lumber-75800593/304600525

- Chicken wire which is sold at Home Depot for $14.62USD for a 24" x 25-foot roll.

www.homedepot.com/p/PEAK-25-ft-L-x-24-in-H-Galvanized-Steel-Hexagonal-Wire-Netting-with-1-in-x-1-in-Mesh-Size-Garden-Fence-3353/315112258

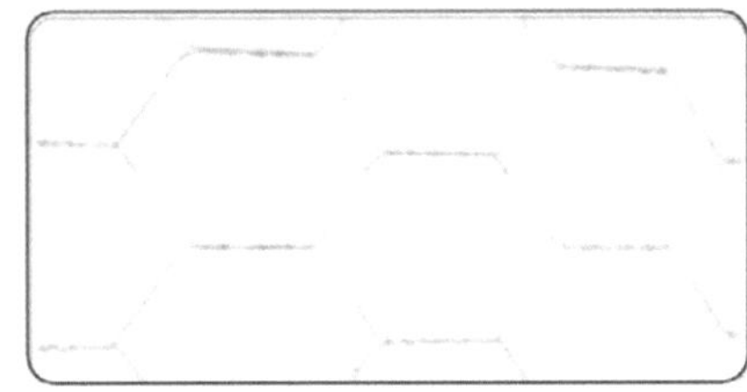

- Five-gallon buckets which are available at Home Depot for $ 4.98 USD.

www.homedepot.com/p/The-Home-Depot-5-Gal-Homer-Bucket-05GLHD2/100087613

- Five-gallon bucket lid which is also available at Home Depot for $ 2.48 USD.

www.homedepot.com/p/The-Home-Depot-5-gal-Orange-Leakproof-Bucket-Lid-with-Gasket-5GLD-ORANGE-LID-for-5GL-HOMER-PAIL/202264044

- A length of rope which I found at Home Depot for $ 5.51 USD for 100 feet.

www.homedepot.com/p/CORDA-1-4-in-x-100-ft-Hollow-Braid-Poly-Rope-PB7703/305347534

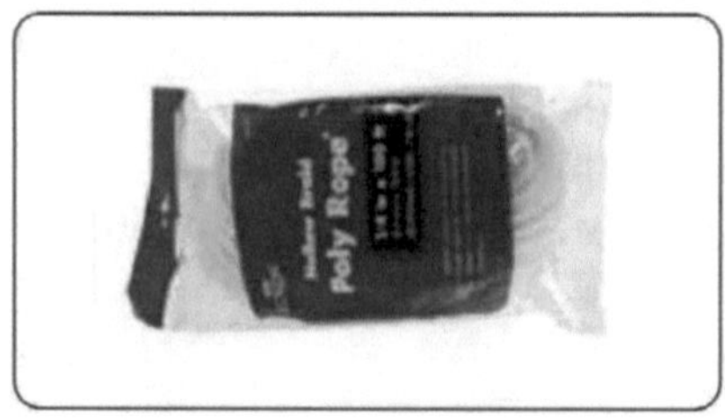

- #8 x 2" Screws which you can find at Home Depot for $ 11.98 USD for a box of 161.

www.homedepot.com/p/SPAX-8-x-2-in-Philips-Square-Drive-Flat-Head-Full-Thread-Yellow-Zinc-Coated-Multi-Material-Screw-161-per-Box-4101020400506/202041005

- Fence staples like these I found at Home Depot for $ 5.98 USD per box.

www.homedepot.com/p/Grip-Rite-3-4-in-Hot-Dip-Galvanized-Staples-1-lb-Pack-34HG-PNS1/100148501

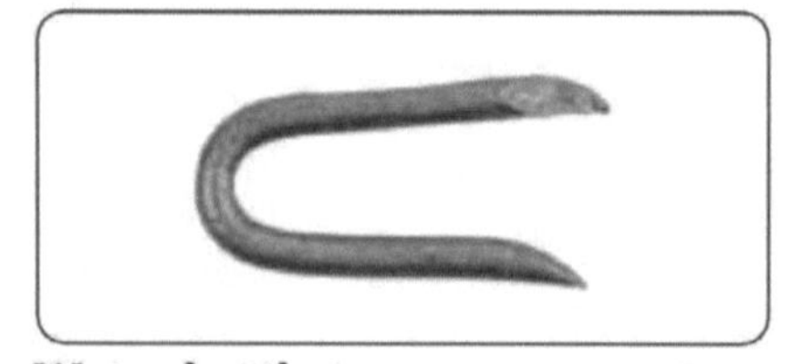

- Water bottle to use as a counterweight.

Instructions

1. Either buy or build a suitably sized cage. The cage's dimensions will be determined by the number of birds you want to catch and the space you have to house the cage. If you are building the enclosure, create a simple frame using the 2x2's after which you will sheet the bottom with plywood and the top with either plywood or another material like foam board. The advantage of using foam board is that it is easier to cut.
2. Wrap the sides of the cage with the chicken wire securing in place with the fencing staples.
3. Cut large openings in the sides of the bucket large enough for a bird to walk through and into the bucket. Secure the lid to the top of the bucket.
4. Use the bucket to layout and cut a hole in the top of the cage. Make sure that the opening is large enough to allow the bucket to pass through it without interference.
5. Construct a simple lever with some 2x2s. Do this by drilling a hole at the end of a 2x2 which you will secure to the corner of the cage. The height of the 2x2 should be high enough that the bucket can be comfortably suspended slightly above the top of the cage.
6. You will secure another 2x2 to the one you just drilled and mounted. One end should be directly over the hole. Mark the point where the two boards intersect and match drill a hole the same size as the first one you drilled.
7. Drill holes at either end of this 2x2 to secure a rope for the counterweight and bucket.
8. Attach a rope to the end over the hole and to the handle of the bucket. Make sure that the bucket will hang into the hole without touching the sides.
9. Attach the counterweight to the other side of the lever. Adjust the weight of the counterweight until it easily lifts an empty bucket from the floor of the cage to above the cage.
10. Place something underneath the counterweight to stop lifting the bucket to a point slightly below the top of the cage.
11. Place bait around and inside the bucket.

Using the Automatic Bird Trap

Birds will not be easily fooled. They will require some time to become comfortable with the situation, but when they do, the trap will be very effective at trapping them alive and unharmed. As the bird steps into the bucket, the bucket will lower into the cage with the bird inside. As long as the bird stays inside the bucket, the counterweight will not return the bucket to the start position.

As the bird steps out of the bucket and into the cage, the counterweight will raise the bucket to the start position leaving the bird trapped in the cage.

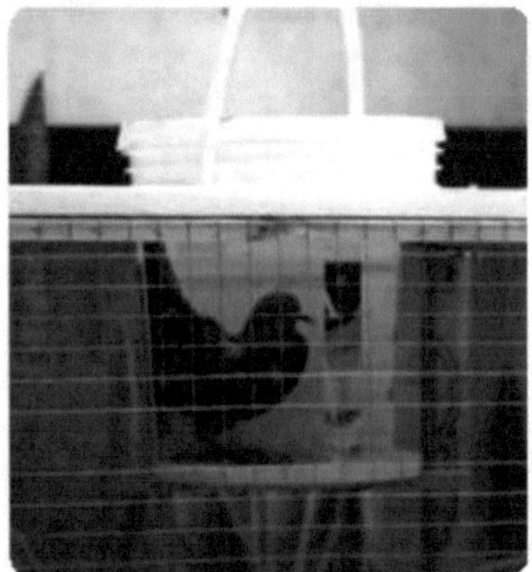

Variations

The automatic bird trap can also have a few variations that are worth noting.

- Installing a door in the side of the cage will make removing the birds a lot easier.
- If you have a coop which you will be housing these wild birds to fatten up for use as food in the future, you can build and install this trap adjacent to this coop to trap and introduce more birds without actually handling them.
- If you leave some birds inside the cage, they may attract others to the trap.

There is an excellent probability that you may have most of the materials around your home or in the surrounding area to construct this trap. Therefore, the automatic bid trap is a fantastic trap to keep the plans in mind for the possibility you find yourself and your family desperate for food.

DIY Mason Jar Soil Test

EASY  **DIFFICULTY** **10 MINUTES** 

Have you ever wondered what the structure of your garden soil is? It would help when making decisions on what and where you might plant, as well as amending the structure of your soil.
This test is so easy that I recommend doing it each year before creating a garden plan.
The results of the test could determine how much fertilizer you use or what changes you should make to the soil, depending on what you plan to grow in your garden.
It comes down to the number of various particles and components that make up your soil. The shapes and sizes of each group of particles determine the soil's structure and how well it will work for growing your garden. A good understanding of the soil and its components will lead to a more successful year of growth.

Understanding the Basic Components of Soil

Most soil contains 3 components: sand, silt, and clay. Each region and area will have a different structure of these 3 components. In fact, different areas within your own property could vary quite a bit from one area to another.

Clay

This is the smallest of the 3 mineral components. It consists of tiny and flat particles that fit together and often creates the biggest area of soil in many areas.
Clay consists of necessary nutrients and works well for storing water. So, having a large amount of clay can work in your favor, unless you have an abundance of rain in a season.
It's also a cooler soil that takes longer to warm up in the spring.

Sand

The largest particle making up soil is sand. These particles are round, which allows more space between particles. As a result, water will drain more quickly from sandy soil than one of more clay.
Another downside is that the nutrients drain right along with it. For gardens with a large amount of sand, it will require more fertilizer and water than soil consisting of more clay.

Silt

The size of silt particles is between clay and sand, but closer to clay. Soil that consists of a large amount of silt is typically found along riverbanks. This soil feels smooth while moist, yet powdery when it's dry.
A good combination of all 3 of these particles is referred to as loam soil. And, it's the best type of soil for gardening. To obtain loam, you need to know what you currently have, so you can make adjustments. So, grab a mason jar!

The Mason Jar Soil Test

Start with a clean and empty jar, such as a mason jar, including the lid.
It could be either quart or pint size. You just need a jar with lid, soil, and water.

1. Fill the jar halfway with soil.
2. Add water. Fill to about 1" from the top and put the lid on.
3. Shake the contents for a few minutes.
4. Set the jar aside for a few hours. As the content settles, it will separate into layers of silt, sand, and clay.

What you should see in a few hours:

- At the bottom, you will see the heaviest layer, consisting of sand and rocks;
- The next layer will consist of silt;
- Next will be the layer of clay.

You should also see some organic matter floating at the water's surface. Also, keep in mind that the color of the soil is indicative of the amount of organic material within it.

The lighter it is, the less there will be. Dark soil holds more organic matter and will warm quicker in the spring.

Reading the Results

Once you have done the test, you need to know what it all means and if you have the desired loam soil.

- Loam – 40% sand, 40% silt, and 20% clay;
- Silty Clay Loam – 60% silt, 30% clay, and 10% sand;
- Sandy Loam – 65% sand, 20% silt, and 15% clay;
- Silty Loam – 65% silt, 20% sand, and 15% clay.

So, knowing that loam soil is ideal, what amendments could you do to achieve it?

Sandy Soil

A good way to amend a sandy soil is to add compost or manure. This is probably the fastest way to get it to a loamy state.

However, both compost and manure contain significant amounts of salt, and high levels of salt can be damaging to plants.

So, if you already have high levels of salt in the soil, such as seaside gardens, use a plant-only based compost.

Silty Soil

You can amend silty soil by tilling in 1" Cocopeat to break up clay particles to help with aeration, 2" of Perlite to help with drainage, and 4" of compost to help with water retention.

Clay Soil

For the same reasons as above with silty soil, till in the same ingredients, but adjust the amount to 1" Cocopeat, 4" of Perlite, and 6" of compost.

My test resulted in about 50% sand, 40% silt, and maybe 10% clay. So, I have fairly good soil in that area. But, to get my perfect soil, I would do a combination of the sandy and silty soil amendments listed above. Or, because it's so close to the desired loam, I could just add fertilizer and water more often.

However, the area I tested is where we normally put a few annuals. We plan on planting a large produce garden on the other side of the house next year, and that will most likely have a completely different test result.

Therefore, it's important to do a different test for every area you intend to plant a garden.

Hopefully, this helps in getting you closer to the perfect soil for your garden. No more excuses! Well, except for maybe pesky critters that you will need to keep at bay.

DIY Survival Garden

No matter if you're planning to start a survival garden out of necessity, prepare for the worst times, or become more self-reliant, you'll love the security these gardens offer.

Homegrown produce boasts better flavors and is more nutritious than grocery store fruits and vegetables. History shows that people have planted survival gardens during the war, economic instability, and famine. Survival gardens assure you that you and your family can easily make it through hard times.

Survival gardens provide access to healthy, readily available food at budget-friendly prices. Do you know how to start growing your survival garden? Here's an easy-to-understand guide to starting your survival garden in your backyard.

Why You Should Grow a Survival Garden

Homeowners often worry about day-to-day shopping for essentials and amenities. Going per week or even monthly can be too troublesome for some people. That's why Survival Gardens are the perfect solution.

The ability to harvest fresh produce from your backyard is an ideal way of filling up your pantry so that you can enjoy delicious and fresh food. On top of that, a survival garden provides access to organic and nutritious veggies and fruits. Typically, these are superior in quality to week-old, conventionally grown store-bought food.

One advantage of survival gardens is that you can grow a diverse variety of delicious veggies that extend the life of the stored food supply. Moreover, gardening is a great way to connect with nature and spend meaningful time with your family.

Plus, gardening helps you relieve stress, as well as ensures you stay fit and healthy. Enjoy breathing fresh air as you enjoy fresh greens rich in starchy carbohydrates and necessary vitamins and minerals.

Design Your Survival Garden

Designing the structure of your survival garden is just as crucial as buying garden seeds. When it comes to deciding what type of garden seeds you should plant, it's always a better idea to seek open-pollinated seeds. These allow you to save your seeds and, ultimately, ensure that you always have stock for the year to come.

Keep in mind that a survival garden is more than simply a garden; it is the thin line keeping you healthy and happy during tough times. Plus, you need to consider more than merely eating crops. From preservation to trading, excess crops are beneficial.

There's always a chance that your crops may fail due to a disease or drought. Thus, you need to learn to think diversely. For this reason, you should plan perennials, as well as annuals. Consider planting trees, canes, bushes, herbs, and weeds. Moreover, add sun and shading loving plants to enjoy delicious veggies regularly.

HOW BIG SHOULD IT BE?

When it comes to how big should your survival garden be, you'll have to consider a variety of different factors such as:

- The number of people in your family
- What kinds of crops will you be growing in your garden- some require more space than others
- What kind and quality of soil does your backyard boast?
- Also, what is your climate?
- Your gardening skills and experience
- The amount of time you must spend on gardening and, in turn, feeding your family.

While it's not possible to provide you with an exact number, you'll likely need a minimum of ¼ acre land for starting your survival garden. Depending on your family size and vegetable preferences, chances are, you may end up with a 2-acre expansive survival garden.

PICK A SITE WITH LIGHT

One additional tip to keep in mind when designing your survival garden is the location. Apart from factoring in the area and size, you'll have to make sure the garden allows adequate sunlight to enter. No matter if you're planting in the ground or pots, your spot should enjoy at least 8 to 12 hours of uninterrupted sunlight per day. A tree's shadow or structure can prevent the sprouting of certain plants.

The fact is that light is critical for the healthy growth of plants. Moreover, most plants are sensitive to interruptions in light. When designing your garden, test the spot from different angles and at a different time to ensure ideal light exposure.

BE MINDFUL OF THE SEASON

No need to grow tropical vegetables in early spring. Instead, focus on frost-tolerant plants. If you're not growing your plants indoors, you'll want to find out the date of your area's 'last frost. The garden elevation plays a significant role in determining frosts, freezes, and nighttime temperatures. Plants native to warm places such as peppers, tomatoes, and eggplants are not cold tolerant. Even the slightest frost will kill them.

That's why you must grow your tropical seedlings during comparatively warmer months or in a warm indoor space. Here's a small guide:

Seeds to Sow During Early Spring

- Greens: lettuce, spinach, swiss chard, and kale
- Root crops: turnips, carrots, radishes, and beets
- Peas: snow peas and early peas

Live plants to Sow during Early Spring

- Onion sets
- Hardy herb: thyme and rosemary
- Cruciferous vegetable seedlings: cabbage, broccoli, and cauliflower

Important Crops for Your Survival Garden

Here's a list of plants to consider growing in your survival garden:

- Corn
- Spinach
- Potatoes
- Onions
- Beans
- Peas
- Tomatoes
- Cucumbers
- Winter Squash
- Lettuce
- Sunflower Seed
- Cilantro.

When growing crops is near impossible during the colder months, you'll have to preserve and store produce. For this reason, you'll have to learn how to cook, keep, and store vegetables for later use. Freezing, dehydrating, and canning are the top couple of ways to preserve your favorite foods for the colder months.

When drought or food insecurity hits you, the first thought that pops into your mind is providing your family with proper food, shelter, and water. That's exactly what a survival garden offers. How- ever, keep in mind that a survival garden isn't all sunshine and rainbows. You may face failure, lim- ited space, limited resources, and difficulty pro- tecting it from critters and plant diseases. How-

ever, keeping your goal at the forefront ensures you're always motivated and level-headed. Allow your dreams to inspire you always to keep moving forward.

DIY Wall-Hanging Herb Garden

MATERIAL COST $ 68.00 MEDIUM  DIFFICULTY 3 HOURS 

Gardening is primarily dependant on the amount of space that you have for your garden. With food security being top of mind, establishing a garden to grow your food is a critical component of off-grid living. The problem is that you will be limited by the amount of available ground available for planting.

The solution is to build a garden vertically, utilizing the unused space above the ground. One of the easiest and best places to do this is to use the backyard fences surrounding most people's backyards.

The location you choose for building a hanging herb garden will depend largely on what herbs you want to plant. Each plant has different sunlight and shade requirements along with soil types and watering. The first step is to formulate a list of herbs and other plants you want to grow as a part of your long-term food security plans. Following this, you must research each herb and determine the precise requirements for them.

After this research, you can plan where on your property would be best for each plant and from there, you can come up with a plan for where you will build a hanging herb garden and what you will be planting in it. You may need to create multiple hanging herb gardens to accommodate each plant.

What to Plant in a Hanging Herb Garden

There are many herbs that you could plant in a hanging herb garden. But, of course, what you decide to plant is largely dependant on your area, tastes, and medical requirements. Still, the following herbs are being included as a starting point in designing your hanging herb garden:

◆ **Cilantro**

- It is a digestive aid and possibly removes heavy metals and other toxins from the body
- Been shown to have anticonvulsant properties, contains antioxidants, and has anti-inflammatory properties
- It also has antimicrobial properties.

◆ **Lemon Balm**

- Antispasmodic effect on the stomach
- It may help reduce stress and treat insomnia
- It has some anti-inflammatory properties.

◆ **Peppermint**

- Relieves indigestion
- It helps to treat colds and flu symptoms
- It can be used to treat headaches and migraines.

◆ **Cayenne**

- Rich in vitamins, minerals, and antioxidants
- Helps with blood clotting
- It can ease cold symptoms
- It might help with pain relief.

◆ **Aloe Vera**

- It helps to accelerate wound healing, especially helpful with burns
- Antioxidant and antibacterial.

◆ **Basil**

- Antioxidant
- It has been used to treat colds, especially inflammation in nasal passageways
- It may reduce high blood sugar.

◆ **Cloves**

- Clove oil is fantastic for treating toothaches
- When used as a mouth rinse, it can reduce inflammation and bacteria.

◆ **Lavender**

- It can improve sleep
- It may be effective at reducing pain
- Lavender can help reduce stress.

◆ **Garlic**

- It may reduce blood pressure and improve cholesterol levels
- Garlic may be effective in treating the common cold
- It has been used for centuries in the treatment of a variety of ailments.

◆ **Ginger**

- It has been shown to help with relieving nausea
- Ginger may help treat cold and flu symptoms
- It may help reduce inflammation
- A good source of antioxidants.

◆ **Sage**

- Contains antioxidants
- Possibly helps control blood sugar.

◆ **Feverfew**

- It helps to treat migraines
- o It May help ease the pain caused by arthritis.

◆ **Parsley**

- It can be used to help with eye health because it is a source of vitamin A
- Freshens breath
- It may help to prevent heart disease
- Parsley could be used to treat UTI's.

◆ **German Chamomile**

- It can be used to treat anxiety
- Often used as a sleep aid when made into tea
- It may have anti-inflammatory properties
- It has been traditionally used to treat a wide variety of conditions.

How to Use Herbs from Your Garden

Each herb that you will plant will be best utilized in specific ways, and it is essential to understand the methods for preparing them for use either medicinally or in recipes. Knowing which parts of the herb to use for different applications is also critical to effectively using a backyard hanging herb garden. There are multiple ways to prepare an herb for medicinal use.

A few examples of preparation methods are:

- ◆ Poultice – A paste made from the herb which is spread on the body using a moist cloth to promote healing.
- ◆ Tincture – A concentrated extract made from the herbs through soaking in alcohol or vine- gar.
- ◆ Infused oil – Are made by infusing a carrier oil with an herb. These can be used in cooking, as the base for salves or balms, in soaps, or di- rectly on the skin.
- ◆ Salves and Balms – Are made by combining herb-infused oil with a wax such as beeswax. These can then be applied directly to the skin.
- ◆ Teas – Are usually made from the dried leaves of your herbs steeped in hot water until you get the desired concentration.

- It may seem daunting to make medicinal herbal products, but with a bit of research and practice, you will be able to treat a wide variety of ailments while also being able to spice up your cooking or baking.

Building the Hanging Herb Garden

The construction of a hanging herb garden is not complicated but will depend on your fencing and the amount of space you have to work with. The dimensions that I have listed here are what worked for me and the area that I had available. You will need to adjust the sizes to suit your space and needs.

Tools

You will need:

- ◆ Staple gun and staples
- ◆ Drill with bits and drivers
- ◆ Tape measure
- ◆ Saw

Materials

As far as materials go, I used:

- ◆ Four pieces of 2 x 8 x 8' pressure treated lumber which is available at Home Depot for $ 12.68 USD.

www.homedepot.com/p/2-in-x-8-in-x-8-ft-2-Ground-Contact-Hem-Fir-Pressure-Treated-Lumber-549000102080800/206931771

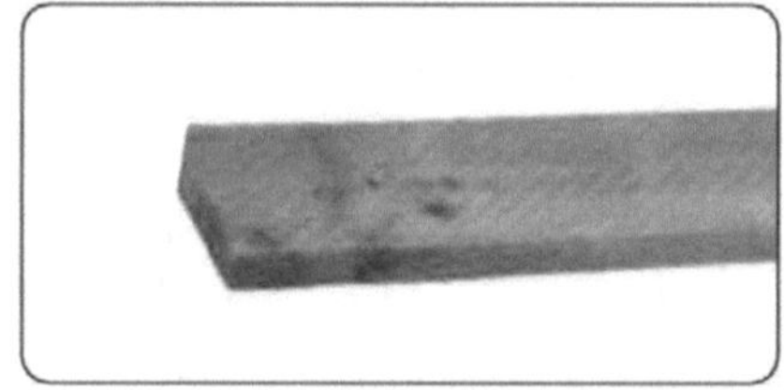

- ◆ 5/8" x 5 ½" x 8' Cedar Fence picket, which you can find at Home Depot for $ 5.24 USD each.

www.homedepot.com/p/Alta-Forest-Products-5-8-in-x-5-1-2-in-x-8-ft-Western-Red-Cedar-Flat-Top-Fence-Picket-63028/205757691

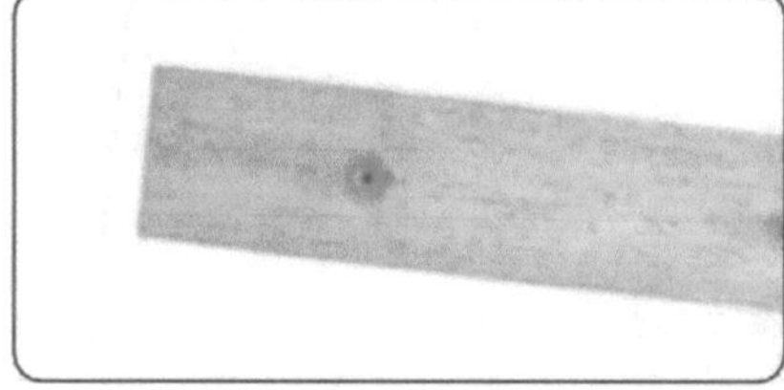

- ◆ A roll of landscape fabric can be found at Home Depot for $ 29.98 USD per 100' roll.

www.homedepot.com/p/Vigoro-4-ft-x-100-ft-Polypropylene-Landscape-Fabric-Weed-Barrier-2239RV/311040977

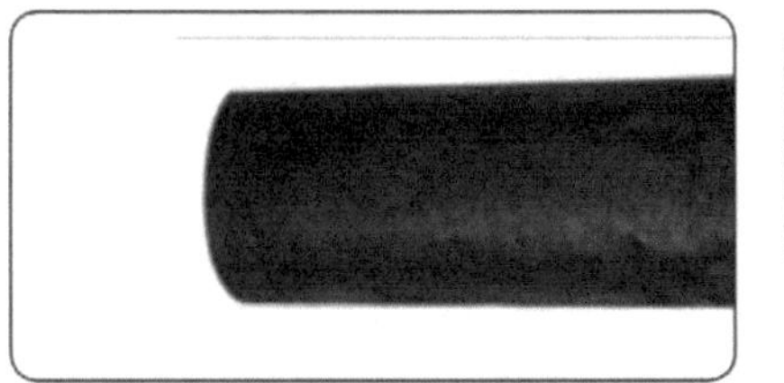

- #8 x 1 ½" Screws which are $ 8.78 USD per pack of 100 at Home Depot.

www.homedepot.com/p/Everbilt-8-x-1-1-2-in-Zinc-Plated-Phillips-Flat-Head-Wood-Screw-100-Pack-801842/204275487

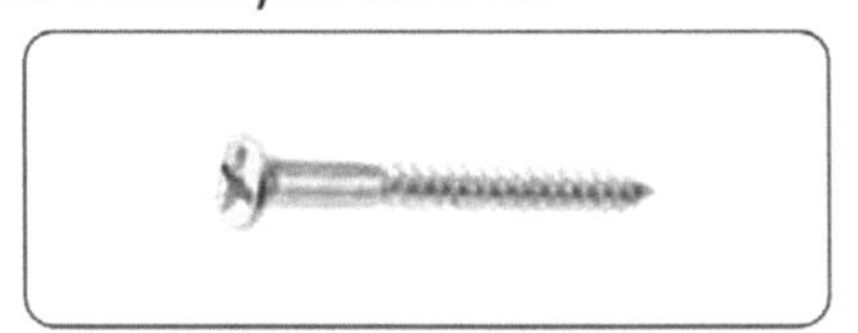

- Potting soil, which is $ 9.97 USD for a 25qt bag at Home Depot.

www.homedepot.com/p/Miracle-Gro-25-qt-Potting-Soil-Mix-72781431/206457033

Instructions

Putting the garden together is a straightforward process. Still, as I stated earlier, the dimensions depend on the space you have available, so you will have to adjust the measurements that you see here for your area. I designed this hanging garden to make the best use of the lumber with minimal crops.

1. Measure the space and determine how you would like the hanging herb garden to be arranged. In my case, I chose one fence panel and arranged the garden beds in a checkerboard fashion.
2. Once you have determined the number of herb garden beds you need, you'll have to figure out how big each bed will be. In my case, I decided to build them 24 inches long by about 6 inches wide with a height of one board width. Depending on what you plan to grow, you may want to build your beds larger or deeper.
3. Cut the lumber for the beds. For the garden beds that I will use, I cut three pieces at 24 inches and two pieces at 6 7/8 inches.
4. If you are using nails, lay one of the 24" boards down and get a few nails started along one of the long edges.
5. Arrange a second 24" board underneath the first, with its edge directly underneath the nails, so they form an 'L' shape.
6. Repeat for the other side so you form a 'U' shape.
7. Nail the end caps on each end.
8. Repeat this for all of the required garden beds.
9. Cut some landscape fabric and line the inside of each of the beds, stapling them in place.
10. Trim the excess.
11. Cut the four pieces of pressure-treated lumber to the height of the fence.
12. Level and screw these pieces to the fence panel, keeping an appropriate spacing to achieve your desired pattern. In my case, I chose a 24-inch centre to centre spacing for each of the boards.
13. Layout where the planters will be mounted by drawing a line with your level at the heights you want the planters' top to be.
14. Mount each bed by drilling pilot holes then screwing them in place.
15. Fill the beds with soil and plant the herbs.

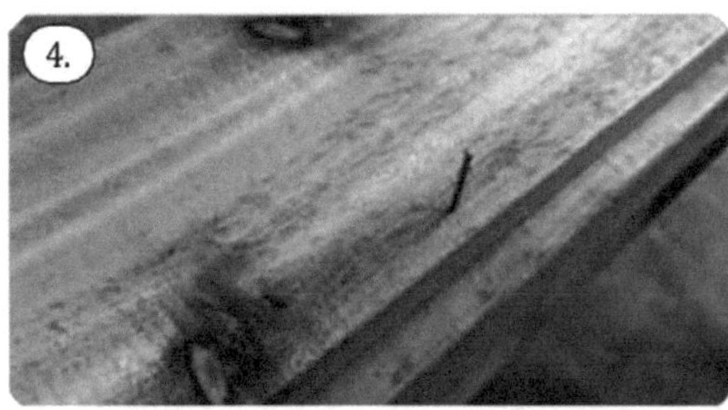

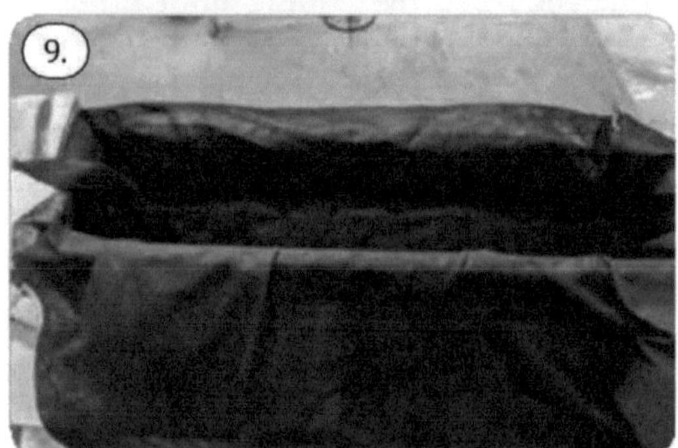

Variations

There are a couple of variations that you could consider employing depending on the situation you find yourself in.

- Instead of installing the hanging herb garden on a fence, you could cut two boards so that they can lean against a fence or wall at an angle. Then, the planters can be screwed into the boards so that they form a ladder.
- The same style of hanging herb garden could be built and used indoors as well. The only issue would be dealing with any water that may be leaking out from the planters. In this instance, plastic planters with no drain holes may be more appropriate.
- Each of these planters can also be covered to form a mini greenhouse in the environmental situation warrants it.

A DIY hanging herb garden is a fantastic way to utilize the vertical space around your backyard, expanding your ability to grow herbs and vegetables, and take a further step towards food self-sufficiency.

How To Make Cabbage Bandages To Treat Inflammation And Joint Pain

EASY DIFFICULTY

When I was a child, I didn't stay indoors much. I used to play with other kids in my neighborhood all day long. Of course, this meant injuries were an almost daily occurrence.

But I didn't care. I was just a kid doing things that every other kid did back then. Whenever I'd come back home, my mother would look at me and sigh, "Christ, not again..."

She would then prepare some cabbage leaves and wrap them around my wounds, using bandages to keep them in place.

After a time the swelling would go away, the bruises would be significantly reduced, and cuts would be almost completely healed. And that happened much faster than normal.

We weren't a wealthy family, so we couldn't afford to go to the doctor for every minor thing.

However, my parents were very knowledgeable about the natural remedies passed on by my grandfather.

For centuries, people all over the world have used cabbage leaves to successfully reduce swelling, pain, and strains.

It is high in vitamins and phytonutrients, as well as anthocyanins and glutamine; both of which have anti-inflammatory properties.

Furthermore, modern science shows that cabbage contains 2.6% to 5.7% sugars, 1.1% to 2.3% proteins, fixed oil, and mineral salts, including sulfur and phosphorus.

The plant also contains vitamin C and S-Methylmethionine, also known as vitamin U, which is antiulcer.

This makes it a very powerful and convenient tool against joint pain, arthritis, and most injuries.

What You'll Need

Here is a list of everything you'll need:

- A cabbage (obviously);
- Bandages;
- Cellophane;
- A cup, hammer or rolling pin (basically anything that gets the job done);
- A cutting board.

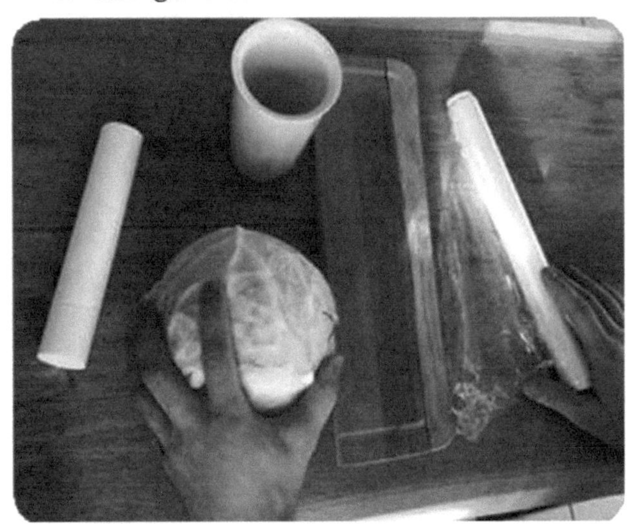

How to Make Cabbage Bandages

1. Place cabbage leaves (green or red) on a cutting board and cut out the hard stem.
2. Hammer the leaves with any kitchen utensil to gently bruise the leaves in order to release some of the cabbage juices.
3. Layer the cabbage leaves around the knee or ankle joint until it is completely encased with the leaves.
4. Hold the leaves in place by wrapping them with bandages.
5. Wrap all of this up with cellophane in order to hold the warmth and cabbage juice around the skin.
6. Leave the cabbage leaves wrapped around the joint for at least one hour.
   If no skin sensitivity is noted, the leaves can be left on overnight.
7. Unwrap the cabbage leaves when cool and discard.

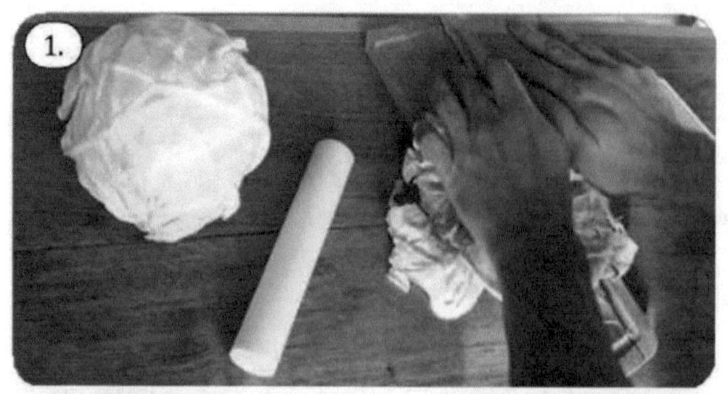

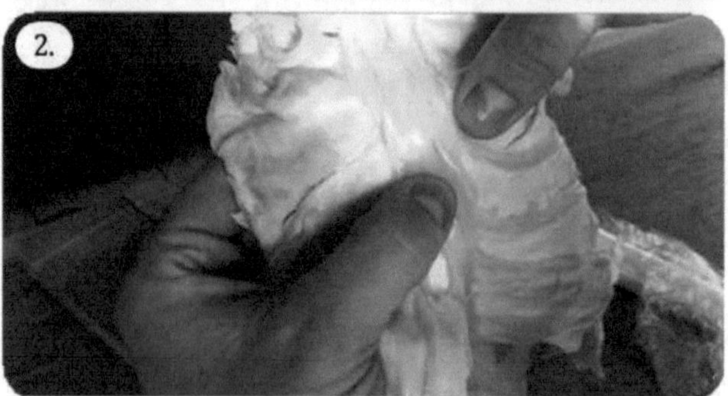

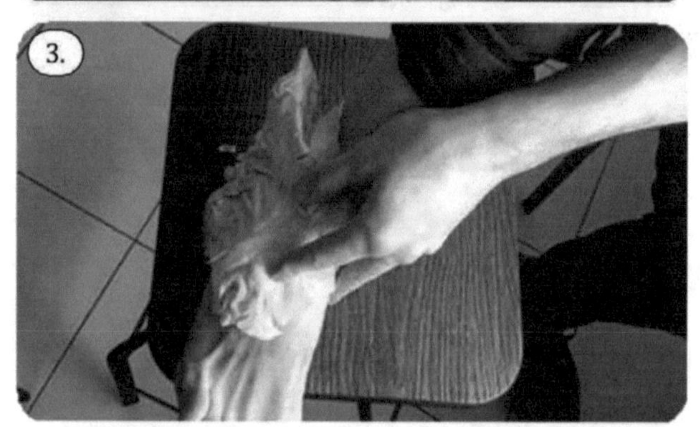

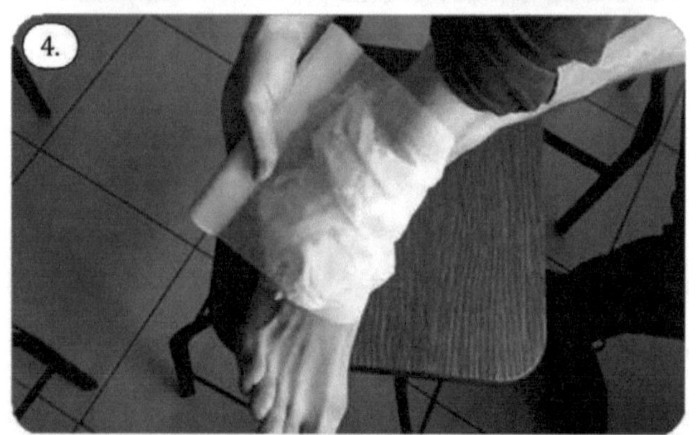

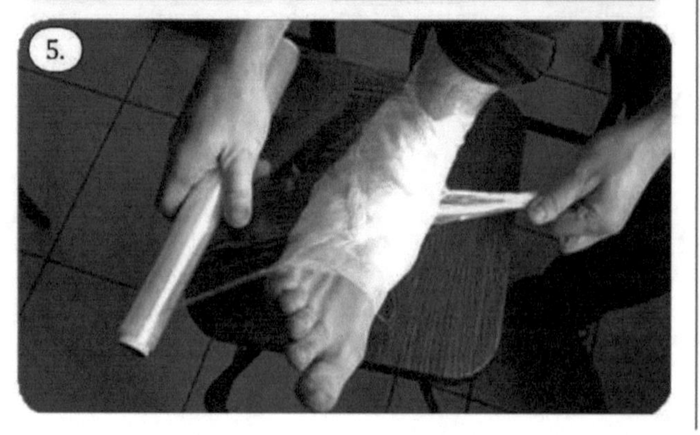

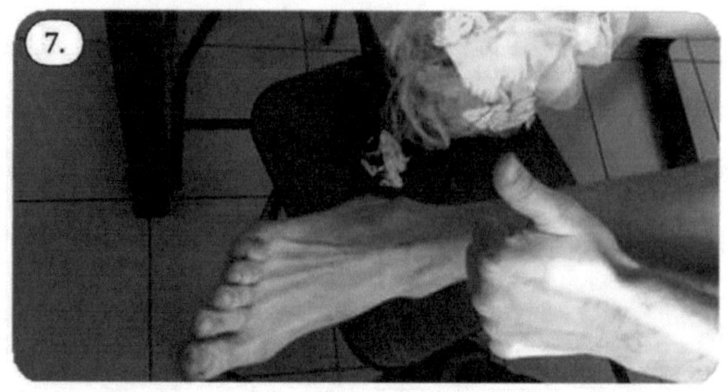

Here are some of the things cabbage bandages can help you out with:

Eczema

Use cabbage leaf bandages for about one hour.

Asthma

Apply four cabbage leaf bandages on the chest or shoulders for at least four hours.

Arthritis

Pound the cabbage leaves with any kitchen utensil you have on hand (even a simple cup) and apply it directly to the affected area.

Wrap it up in a bandage and cellophane to make sure it stays tight so that the skin absorbs the vital nutrients.

Leave it on for several hours and repeat this process until you see a huge improvement.

With all of these benefits, how can you not love cabbage? God has indeed blessed us and we can gain so much simply by studying his creation.

How To Make Moringa Powder

EASY  DIFFICULTY

Numerous studies have already been conducted regarding the medicinal properties of moringa and it has been found to be beneficial for many conditions, ranging from skin diseases to hypertension, diabetes, kidney stones, tuberculosis, and even tumors.

In Ayurvedic medicine, moringa is cited to have the ability to treat more than 300 illnesses and diseases.

Aside from vitamins, moringa is also rich in minerals, antioxidants, and antibacterial and tissue protective properties.

The number of nutrients it contains is staggering: 92 nutrients, 46 antioxidants, 36 anti-inflammatories, 18 amino acids, and 9 essential amino acids!

So what does that mean for you? It means if you choose to incorporate moringa into your daily diet, your body will be able to:

- **Fight Free Radicals**

Free Radicals cause oxidative stress and cell damage. By fighting these, your organs will stay healthy and function optimally.

- **Fight Inflammation**

Moringa helps treating chronic diseases such as diabetes, arthritis, respiratory problems, and cardiovascular diseases while even reducing or avoiding obesity.

- **Protect Your Brain from Alzheimer's**

Moringa's antioxidants and neuro-enhancers improve cognitive function and support brain health.

- **Ward Off and Fight Infections**

Moringa's natural antimicrobial and antibacterial properties are effective against a host of microbes, bacteria, and fungi that are responsible for all kinds of infections.

- **Protect Your Liver**

Moringa's high concentrations of polyphenols and other antioxidants can protect your liver against toxins and oxidative damage.

- **Keep Your Skin Youthful and Radiant**

The antioxidants in moringa not only fight toxins and free radicals, but also shield your cells and tissues.

- **Improve Your General Health**

Indeed, nothing comes close to moringa when it comes to providing your body with nutrition, health, and beauty.

The most convenient and enjoyable way to take moringa is by its powder form.

Step-by-Step Guide on How to Make Moringa Powder

1. Harvest

If you have moringa trees in your backyard, simply harvest a bunch of stalks (about two pounds). You can also buy some at the market. Always opt for the mature, rich green leaves.

2. Sanitize

Sanitize a basin, pan, bowl, or any vessel that you can use to wash the leaves.

Sprinkle a few tablespoons of baking soda into the water to clean the dust and other impurities off the moringa leaves.

3. Cleanse

Wash the moringa leaves, removing dead and yellow leaves and any infected parts.

4. Hang

Shake the excess water off the leaves, tie the ends of the stalks together, and hang them upside down in an enclosed place that doesn't get direct sunlight, to preserve the nutrients.

5. Air Dry

Leave the leaves hanging for three to four days until they are brittle to touch.

6. Separate

Separate the leaves from the stalks and stems. The fewer stems there are, the smoother the powder will be.

7. Grind

You can use a blender to grind the moringa leaves into powder form.

Run from 30 seconds to a full minute or till you achieve the desired texture.

8. Store

Keep the moringa powder in an airtight container to preserve the nutrients and store it in a cool, dry place.Keep the container closed at all times to keep moisture out and to preserve a longer shelf life. The powder will last up to six months without preservatives.

Now you're ready to try your moringa powder and experiment with a myriad of ways to enjoy it! You can make hot tea, cold tea, or iced tea out of it, or you can add it to your smoothies, shakes, and salads.

To make moringa tea, just add a teaspoon of moringa powder to hot water. You can also add peppermint leaves and lemon for flavor and sugar or honey to taste.

Here are some other ways to enjoy your moringa powder:

Mangosteen Moringa Tea

A combination of mangosteen tea and moringa powder will yield a very strong antioxidant and antibacterial concoction. Mangosteen is an exclusive source of xanthones, which can inhibit cancer cell growth.

Simply add a teaspoon of moringa powder, along with some lemon and honey, to mangosteen tea, and you will have a very refreshing and healthy beverage.

Moringa Chicken Soup

What you will need:

- ½ lb. cut chicken;
- 2/3 small slices of Fresh Ginger;
- Lemongrass leaves;
- Bell pepper;
- Table salt;
- 1 Tbsp. moringa powder.

1. Sautee the cut chicken on a few slices of ginger.
2. Add a cup or two of water and salt to taste; throw in the bell pepper and a few lemongrass leaves.
3. Bring to a boil until chicken is tender, then sprinkle moringa powder.

Moringa Fish Soup

What you will need:

- 1/2 lb. fish;
- 2/3 small slices of Fresh Ginger;
- 1/2 cup coconut milk;
- 1 bell pepper;
- 1 tomato;
- 1 onion;
- Table salt;
- Moringa powder.

1. Fry fish and set aside.
2. Slice the onion, tomato, bell pepper, and ginger.
3. Boil two cups of water.
4. Add onion, ginger, bell pepper, fried fish, and salt to taste.
5. Boil for ten minutes more then add moringa powder and coconut milk.
6. Boil for 15 seconds more, then serve. Never boil coconut milk longer than 15 seconds or it will start to turn into oil.

Moringa Lentil Curry Soup

What you will need:

- 2 teaspoons edible oil;
- 1/4 lb. pork, in cubes;
- 1 medium onion, diced;
- 1/2 Tbsp. grated fresh ginger;
- 1 Tbsp. curry powder;
- 1 cup baby carrots, grated;
- 1 cup lentil;
- salt to taste;
- a pinch of black pepper powder;
- 1 cup coconut milk.

1. Soak the lentil in some water an hour or two before cooking to reduce cooking time.
2. Preheat pan, then add oil.
3. Sautee onion till golden brown.
4. Add ginger and curry powder and stir for 30 seconds.
5. Add pork and salt to taste, stir for one to two minutes.
6. Add just a little water (about 1/4 cup) to keep the meat from being burnt.
7. Cook in slow fire, till meat is tender.
8. Add 2 cups water, bring to a boil.
9. Add lentils, more salt if needed, and pepper. Keep adding a little water as necessary.
10. When lentil is soft, add carrots and moringa powder.
11. Simmer for 2 minutes, then add coconut milk, then simmer for 30 seconds.

How To Make Your Own SHTF Medicinal Garden

There are several ways you can start your SHTF garden. It all depends on how much money you want to spend, how much space you have to work with, and how much work you want to put into it. The first thing you should do is assess your property and determine a good spot to place your garden. You will need a spot for plants that require full sun, plants that require partial sun, a few medicinal bushes, and plants that require shade. You should also look for a spot to plant invasive plants so you don't compromise other important plants in your garden. If you are planning on a food garden, make sure you leave enough room for this as well.

How Much Space Do You Need?

You do not need a huge yard or farm to create an effective and useful medicinal plant garden with the plants discussed further in this chapter. Building a few raised beds that are roughly 4x12 (48 square feet) will suffice. This can easily be accomplished on plots of land no bigger than ⅙ to ⅛ of an acre. With that much space, you can have several large raised beds in various areas of sun, while also allowing a spot for the medicinal bushes and potentially invasive medicinal plants to thrive.

Materials Required

While it is entirely possible to create a SHTF garden by tilling up areas of your yard and sowing seeds/planting seedlings directly in the soil, your best bet for the survival of your potentially life-saving plants is in raised beds. As you are probably already aware, all gardens need weeding. Weeding gardens can take up a huge amount of time and leave you down in your back if you aren't careful. Putting in a little extra work at the beginning of this endeavor will save you countless hours weeding in the future!

Consider saving money by chopping down cedars to place on the ground for your raised beds. Cedar is an excellent choice because it is a natural pest repellent. When you chop the cedars down, trim off all the limbs and dig a small groove in the ground to lay the logs flat. Trim the logs to be somewhat flush as you lay them down in your desired areas. You can either buy dirt to fill the raised beds with, or you can use a tractor to move dirt from another area of your property to the raised beds.

If you have livestock, consider mixing in manure with the soil for a natural fertilizer. Horse manure is optimal, but you can also use chicken, sheep, cow, or other types of manure. Don't go overboard, however, or you can burn up your crops. A little will go a long way.

Make sure you have a good place indoors for that gets plenty of sunlight. Perhaps a sunroom or an area with several windows would be the perfect spot. Set up a table to place your seedlings on and keep them moist and warm so they can thrive.

Another option, if you have the money is to build a small greenhouse outdoors near your garden area. This makes it much more convenient when you want to move some of your seedlings outside but don't want to plant them in the garden right away. If you have heat to your greenhouse, you can also start your seedlings in there. Some peo- ple heat their outdoor greenhouses with wood heat; just make sure you have good ventilation. Again, using cedar along the bottom and for the beams can help repel pests in your greenhouse. Keep in mind that a greenhouse is preferred, but not required.

Access to a water source is important. Make sure you have a spigot or a source of water nearby. If you are off-grid and do not use electricity, consid- er placing a rainwater collection barrel near your garden so you can water plants easily.

Gardening tools like a spade or trowel for dig- ging small areas to plant seedlings, small trays for planting seeds, a device for tilling soil, a hoe, a rake, a pruner for your bushes, and scissors for trimming and harvesting herbs are all very handy to have in this endeavor. If you are off-grid, a chainsaw is almost essential for chopping down cedars and being able to make your own raised beds without purchasing lumber. A way to drag or haul the cedars is also important. There are sev- eral ways to do this: you could use a mule or two if you have them, a tractor, a truck, or a utility ve- hicle/four-wheeler. It all depends on how far you need to haul the cedars and how independent you are.

Handy but optional tools like a moisture con- tent reader, weeding tool, seed distributor, and gas-powered tiller are also something to consid- er, depending on the funds you have set aside for your garden.

Pest control is an issue in any garden. There are several things you can do to stay on top of this problem, such as using cedar for your raised beds or bordering your garden with plants that repel insects naturally, like lavender and peppermint (they serve a dual purpose, as they are medicinal as well).

Diatomaceous earth is a natural and safe alter- native to commercial insecticides, which are not recommended for use in a medicinal plant garden. Another natural pest control method is neem oil. Simply add a few drops to a spray bottle of water and spray your plants as needed.

Plants Required

The best way to create a true SHTF medicinal gar- den is to think about all the different systems of the body and try to choose plants that help with conditions for these systems. You never know what kinds of issues you will run into in a SHTF situation, so you want to be ready for anything. Choose plants that target these systems specifical- ly and help treat common ailments of these sys- tems. Below are systems of the body and plants you can grow to target ailments of each system:

1. Medicinal Plants to Have at Hand for Your Circulatory System

Common issues with the circulatory system in- clude high blood pressure and blocked arteries from high cholesterol. If these remain untreated (and in a SHTF scenario these will likely get worse) damage to the heart, or even death, can occur.

a. Lavender (*Lavandula angustifolia*) for blood pressure

A great plant to grow for treating high blood pres- sure is lavender. It is such a low-maintenance plant once it is established! It does not require much watering and does best in full to partial sun. It grows in a variety of growing zones, but does best in USDA zones 5-9. It prefers well drained soil. Avoid planting lavender in acidic soils. Try planting lavender seeds indoors in early spring and achieving a strong seedling before planting in your garden. Lavender has been proven through scientific research to lower blood pressure and help promote calm and peace.

To prepare lavender for medicinal use, harvest by clipping the plants in full bloom. Next, chop the plants finely and place the chopped plant matter in a glass jar. Cover the plant material completely with at least 80 proof alcohol and let this infuse for four to six weeks, shaking daily. When it is fully infused, strain out and bottle the liquid. Take one to two 5 ml doses daily to help manage blood pressure issues. You may increase or decrease as needed.

b. **Garlic for cholesterol**

Common garlic has been shown to lower cholesterol. It is an easy-to-grow plant that many gardeners include in their gardens. It is a wonderful addition to any meal! Garlic can be grown in almost all growing zones. Since garlic is so common, it can easily be grown by breaking the cloves out of a bulb of garlic and planting them in well-draining soil about two inches down and eight inches apart. Make sure when you are breaking the bulb into cloves that you try to leave as much of the papery substance that surrounds each clove intact. The best time to plant is said to be in the fall, around six weeks before the ground freezes. Cover the cloves with a thin layer of compost and fertilizer. Water them daily or as needed, depending on the weather and the dryness of the soil. If you planted in the fall, the most common time to harvest is in late July.

The best way to use garlic is to use it as often as possible in meal preparation. If you don't think you are getting enough garlic in your diet by adding minced cloves to meals, you can chop up cloves and add them to a jar of raw honey and let it ferment. Honey-fermented garlic is highly medicinal. It is a great way to treat a variety of infections and a great way to keep cholesterol in check.

2. Medicinal Plants to Have at Hand for Your Digestive System

Common complaints of the digestive system include gas/bloating, diarrhea, and constipation. While gas and bloating are usually minor, diarrhea can be life-threatening if it is not treated soon. Below are plants you can grow to manage these issues.

a. **Peppermint (*Mentha piperita*) for gas/bloating, upset stomach, nausea, and indigestion**

Peppermint has long been utilized for its ability to soothe a variety of digestive woes. One of the best things about peppermint is how easy it is to grow! You can start peppermint seeds indoors in early spring by shaking a few into separate trays. When they have sprouted and are strong enough to transplant outdoors, you can plant them (after the first frost) in your raised bed, or pretty much anywhere on your property. Water them as needed and watch them take over the area you planted them. Make sure to keep an eye on them, as they can spread fast and take over other plants. Peppermint can grow in almost all USDA zones. Harvest peppermint by clipping as needed and drying the plants (laying on a drying rack or hanging in small bundles will work).

When it is dried, use the dried leaves to infuse in a cup of hot water (using reusable tea bags or a tea infusion ball) and drink as needed to relieve stomach issues.

b. **Agrimony (*Agrimonia eupatoria*) for diarrhea**

Agrimony is a gentle and effective way to help calm the stomach and treat diarrhea. The first thing to do if you have diarrhea is to make sure you are drinking more water than normal. You want to avoid dehydration, which can come on quick if you have diarrhea.

Agrimony is native to many areas in North America, but it can also be grown in a medicinal garden if you don't have the time to go looking for it. Agrimony is best grown in USDA zones 6-9. Plant agrimony seeds indoors in early spring and then transplant seedlings in your garden after any danger of frost has passed. Make sure then are planted around twelve inches apart in well-draining soil. Clip the above ground parts of the plant when it is mature and dry them to make tea to help treat diarrhea. Drink peppermint tea as well to further calm the stomach.

c. **Licorice root (*Glycyrrhiza glabra*) for constipation, heartburn, and ulcers**

Licorice is surprisingly tolerant of a variety of conditions, but grows best when it is watered regularly and planted in well-draining soil. It prefers full sun and optimal growing zones are USDA zones 9-11. Plant seeds indoors and transplant to optimal conditions in your garden when they are well-established. Harvest the full plant when it is mature.

The root is what is needed for medicinal purposes. Before harvesting, you may want to clip a cutting off the plant and put it in water in the sunlight, as it can grow new plants from cuttings! Chop the roots well and dry them to drink in tea to relieve chronic constipation.

3. Medicinal Plants to Have at Hand for Your Endocrine System

The thyroid seems to be one of the most affected organs of the endocrine system. It is highly susceptible to environmental toxins. Issues with the pituitary and ovaries are also common. Adaptogenic herbs like ashwagandha are perfect for restoring balance to all parts of the endocrine system, but especially the thyroid. For the pituitary and ovaries, Vitex is extremely helpful.

a. **Ashwagandha (*Withania somnifera*) for thyroid support and overall endocrine health**

Ashwagandha is what herbalists refer to as an adaptogen. Adaptogens are amazing plants because they can do what many plants cannot: target what needs fixed in the body and restore balance/health. Ashwagandha is known for its ability to calm how the body reacts to stress. As a result, the endocrine system is greatly affected in a good way. It can be cultivated in a medicinal garden for its powerful roots. It can be grown as an annual in USDA zones 3-10, but it does best in warmer, drier zones. It prefers growing in temperatures between 70 and 90 degrees Fahrenheit. It also prefers sandy, rocky soil and full sun. Start seeds indoors and plant strong seedlings in your garden (with suitable conditions) after the first frost. Water as needed, but try not to overwater. If you live in a cooler zone, you may want to consider growing this plant indoors in a large pot.

When the plant is fully mature, harvest its roots, which will have a smell described as "sweaty." Chop them well and place them in a glass jar. Completely cover the chopped roots with at least 80 proof alcohol. Let this sit and infuse for four to six weeks, shaking the jar (lid closed) daily. Strain out the extract when it is ready and bottle the liquid for medicinal use. Take 2.5 mil twice daily for thyroid and endocrine support.

b. Vitex (*Vitex agnus-castus*) for hormonal balance and restoration

Vitex, or chaste tree berry, is also an adaptogen. It is known for restoring balance to hormones and helping treat conditions of the pituitary and ovaries. For the pituitary, it can restore normal hormone levels, reducing prolactin levels as needed. For the ovaries, it can reduce symptoms of Poly Cystic Ovarian Syndrome. This is not an herb, but rather a bush. Plant this in an area of your property that would accommodate a medium-sized bush, and not in your garden bed. The easiest way to grow a chaste tree bush is to grab a cutting. They grow roots rather easily when submerged in water for a while. Once you have good suckers established on your cutting, plant the shrub in a large pot with well-draining soil and water it as needed. When it is established well, transplant it (after the last frost) to a part of your property that has full to partial sun. Make sure the soil drains well. Fertilizing this bush every few years is recommended. The berries can be harvested to make a medicinal preparation. Collect berries and place them in a jar. Fill the jar with at least 80 proof alcohol, making sure the berries are fully submerged. Place a lid on the jar and store it in a cool, dark place for four to six weeks, shaking it daily. Strain out the liquid when the time is ready and store your extract in a bottle. Take 2.5 ml twice daily to manage symptoms of PCOS or other hormonal and pituitary issues.

4. Medicinal Plants to Have at Hand for Your Integumentary System

Your skin is your largest organ-and your first line of defense when it comes to foreign objects and bacteria entering your body and causing destruction! Some common issues of the skin are rashes/ irritation and wounds. There are several plants you can use to promote wound healing and calm a variety of rashes and irritations.

a. Lavender (*Lavandula angustifolia*) for calming redness, irritation, burns, and rashes:

Lavender is one of those multipurpose plants you can use for all kinds of issues. It has already been discussed for use in calming anxiety and lowering blood pressure. However, this amazing plant has been used for centuries for calming inflamed skin. Whether the inflammation and redness is caused by a rash, chapped skin, or a burn, lavender has you covered. Growing instructions have already been provided under the circulatory system section of this chapter, but using lavender for skin issues requires a different preparation after it is harvested. First, harvest the aerial parts when the plant is in full bloom. Next, hang or lay your lavender flat on a screen to dry. When it is dried, crumble it into a glass container. Finally, cover the plant material completely in olive oil (or a different skin-nourishing oil like jojoba, rose hip, coconut, etc.). Let this sit and infuse for four weeks, shaking the jar daily. Strain the oil out when it is ready and apply this to irritated skin, burns, or chapped skin to heal it.

b. Yarrow (*Achillea millefolium*) for cuts and scrapes

Yarrow is known for its ability to help the blood clot, as well as its ability to cleanse a minor wound. This comes in handy in a SHTF situation where it is important that a wound does not get infected. While yarrow can be found growing in the wild in most of the United States, it can also be cultivated in a medicinal garden. Make sure you get seeds from the white variety of yarrow, *Achillea millefolium*. To grow yarrow, plant seeds indoors in early spring. They need plenty of light to germinate, and sometimes they may take a while to germinate. Give them patience and anywhere from two weeks to three months. Transplant the seedlings to your garden as soon as they sprout. They require loamy soil that drains well. Yarrow grows best in USDA zones 3-9. They do well in full sun and do not require a lot of water once they are established. You may need to trim them back every now and then. To use medicinally, harvest the aerial parts of the plant and dry them. For minor cuts and scrapes you can apply the yarrow straight to the area after mashing it into a poultice with your fingers. Leave it on the area until the bleeding stops. You can also harvest yarrow and dry the aerial parts to grind into a powder that acts as a styptic powder for wounds.

5. Medicinal Plants to Have at Hand for Your Immune System

There are many reasons for a SHTF situation, but one obvious reason is a pandemic. The whole world knows about Covid and how it spread quickly, sparing some and killing others. Before you find yourself worrying about any virus, keep in mind that most of the battle is keeping yourself healthy so your immune system can fight battles like it is supposed to. This requires keeping it healthy with a balanced, healthy diet first and foremost. As the saying goes, "the best defense is a good offense." Another thing you can do to keep your immune system strong and ready for anything is to grow these plants in your medicinal garden.

a. **Astragalus for preventing overactive immune response and nourishing the immune system**

One big reason why so many people succumb to a virus (other than serious underlying health conditions) is an overactive immune response to the virus. Astragalus can help keep the immune system strong, while balancing its response to unwelcome viruses and pathogens. Astragalus is an adaptogenic plant, so it can help keep your immune system working like it should, despite what your body is confronted with. There are many species of astragalus, but the Chinese version, *Astragalus membranaceus*, is best. It prefers full sun and soil that drains well. Start seeds inside by first rubbing them gently with sandpaper to scrape off some of that tough outer membrane that may prevent germination. Don't scrape too hard because you can damage the seed. Next, soak the seeds in water overnight. The next day they should look swollen. There may be some that are not swollen. To make sure they germinate, poke them with a needle lightly without disturbing the inner portion of the seed. Plant your seeds in small containers indoors in a 2:1 mixture of soil and sand. Keep the soil moist, but don't over water. Transplant the seedlings to a bigger pot with the same soil/sand ratio as the plants grow. They can be planted outdoors in early spring, after the last frost. Be careful with their roots, as they can be tender. Do not over water your plants; just make sure the soil stays moist. USDA zones 6-9 are optimal for growing this plant. It will take around three to four years for your astragalus roots to be big enough to harvest for medicinal use.

When it is ready, harvest the root, wash it off, and chop it into pieces. Place the pieces in a glass jar and cover them completely with at least 80 proof alcohol. Let this sit and infuse for four to six weeks, shaking it daily. Store it in a cool, dark place during this time. Strain out the liquid at the end of the infusion period and bottle it. Take 2-5 ml up to three times daily if you have a virus. For maintenance, take 2-5 ml once daily.

b. **Chinese Skullcap (*Scutellaria baicalensis*) for killing viruses**

Another way to stay on top if you have a virus is to target the virus itself and not just the immune system. Chinese skullcap is antiviral and can be of great help when taken in conjunction with astragalus when you are ill. This Chinese relative of native US skullcap species can grow in North America quite readily. It requires full sun in USDA zones six or higher. It can tolerate partial sun in zones 7-8. It does best when started indoors and can be transplanted in your medicinal garden after the last frost. It does not tolerate clay soils and prefers well-draining soil.

Harvest aerial parts when the plant is mature. Chop them up well and place them in a glass jar. Cover the plant material with 80 proof or greater alcohol. Let this infuse for four to six weeks before straining the liquid out. Take 2.5-5 ml up to twice daily to target a virus.

c. **Elderberry (*Sambucus* spp.) for antiviral and immune support**

Elderberry has long been used for its ability to nourish the immune system and fight viruses. It is native to many parts of North America, but can easily be grown from cuttings as well. If you wish to grow elderberry in your garden, find a bush and trim cuttings while the bush is dormant. In many states, the local conservation agency actually offers elderberry plants each spring that they grow in their native plant nurseries. This is another option if you wish to avoid looking for an elderberry bush to get a cutting. Make sure your cuttings are slanted as you gather them. Pull off the bottom two leaf buds with your fingers. Some people sit the bottoms of the stems in willow tea for six hours before placing them in water to root because willow tea contains constituents that may act as a root hormone. When you place your cuttings in water, it usually takes around six weeks to see suckers if you sit your plants in the sun. Once they have grown nice roots, transplant them to well-draining soil directly in your garden or in a bigger pot. It is probably best to try planting them somewhere in your garden rather than indoors. These are bushes and can get rather big. Choose a place in your garden that is in full or partial sun. Make sure not to place it too close to your other plants because it will likely grow to a place where it blocks out the sun for them.

Harvest the berries in early fall. You have several options with what to do with them for a medicinal preparation, but the best way for a SHTF situation is to make a tincture (the method already described that entails soaking the plant in alcohol). Tinctures have a better shelf life than other preparations. Dry the berries first by sitting them on a screen (still attached to the stems) in a well-ventilated area with great air flow. Once they are dry, place them in a glass jar and cover them with at least 80 proof alcohol. Let this infuse for four to six weeks, shaking them daily. Strain them out and bottle the deep magenta liquid to take when you want to fight a virus. Take 5 ml up to twice daily for this purpose.

6. Medicinal Plants to Have at Hand for Your Muscular System

Common ailments of the muscular system include pulled muscles and muscle pain from overwork. In a SHTF situation, this is likely to occur because you will be working harder outdoors. Be sure to do your best to be mindful of your body and don't attempt to lift heavy objects by yourself! Below are plants you can grow in your garden to treat muscle aches and pains.

a. **Cayenne for blocking pain at the site**

Growing cayenne in your garden has double the benefits because it is already a common fixture in many gardens for food. Cayenne contains a constituent that can help to block pain when applied externally. Growing cayenne isn't hard either. Start seeds indoors in early spring and transplant them to well-draining soil in your garden when there is no danger of frost. Water them as needed. Not surprisingly, cayennes are used to hot conditions and prefer heat from the sun to grow. Plant them in a sunny area of your garden. Cayenne does not tolerate high nitrogen levels in the soil. Cayenne do best in USDA zones 8 and above. In zones below 7, you may want to grow cayenne in pots indoors, as it requires heat to grow.

To use cayenne medicinally, harvest them when the peppers are mature and red. Dry them and grind them down. Infuse a tablespoon of flakes into a cup of olive oil in a double boiler for up to five hours, or you can combine the cayenne and oil in a glass jar and sit it on the "warm" spot on your stove (if your stove has this option) all day. Strain the flakes out and bottle the oil infusion. Rub this into sore muscles liberally as needed. Avoid rub- bing it onto open skin. Avoid touching your mu- cus membranes after application and make sure to wash your hands well.

b. Rosemary (*Rosmarinus of�icinalis*) for muscle pain, muscle relaxation, and inflammation reduction

How fortunate that one of the world's most common garden herbs is useful as medicine in addition to flavoring food! Rosemary packs quite a medicinal punch and has been used for a variety of ailments. It contains a constituent that can bring relief to sore muscles, as well as reduce inflammation if you have pulled a muscle. Rosemary is grown in gardens across the globe and is generally easy to grow. It does grow best in USDA zones 9 or higher, where it will be an evergreen. In all other zones, it will likely come back each year as the temperatures warm up. Rosemary seeds can be sowed outdoors, but to give them their best chance at germination, many gardeners start them indoors first. They seem to take longer to germinate than other herbs. Plant your seeds up to four months before the growing season starts. After danger of frost has passed, plant your rosemary seedlings in an area of your garden that gets full sun. Since rosemary is native to the Mediterranean region, it is used to dry conditions. Do not overwater this plant. Make sure it is planted in well-draining soil with little risk of standing water.

To use this plant medicinally, clip some and hang it to dry somewhere. When it is sufficiently dried, crumble it up and place what you have in a glass jar. Cover the plant material with a carrier oil like olive oil. Let this infuse in the jar for four weeks, shaking it daily, before straining it out. Apply this oil infusion to pulled and sore muscles as needed. Use with cayenne oil for expedited results.

7. Medicinal Plants to Have at Hand for Your Nervous System

Ailments of the nervous system could involve stress, anxiety, and depression, since these things are the result of the activation of the sympathetic nervous system. It could also involve nerve issues that cause pain in certain regions of the body. There are plants you can have on hand and ready to use should you experience any issues of the nervous system:

a. Valerian (*Valeriana of�icinalis*) for anxiety and stress

Valerian may be more famous for its ability to help you get to sleep, but because it directly affects the central nervous system, it can provide positive benefits for those dealing with stress and anxiety as well. This plant does best in USDA zones 4-9 as a perennial. It is somewhat hardy, and will usually grow well in most gardens. It prefers full to partial sun and well-draining soil. Start seeds indoors in early spring, or sow seeds in an area of your garden that meets its sun needs. If sowing directly in your garden, sow seeds after all danger of frost has passed. Plant seeds around ½ inch deep in the soil and thin plants out to eighteen inches apart when they begin to grow well. The roots are the medicinal part of this plant. Try letting your plants grow a few seasons before harvesting the roots. They will have a distinctive smell that people usually describe as "sweaty feet." Don't let this deter you, though!

This plant contains potent medicine! fter harvesting the roots, chop them well and fill a glass jar. Cover the root material completely in at least 80 proof alcohol and let this infuse in a cool, dark, place for four to six weeks before straining out the liquid to bottle. Take 2.5 ml up to two times daily for anxiety, or 5 ml one hour before bedtime if you want to fall asleep fast.

b. **St. John's Wort (*Hypericum perforatum*) for nerve pain and mild depression**

St. John's Wort grows wild over most of North America, so it is no surprise that it is an easy-to-grow garden plant with all kinds of medicinal benefits. USDA zones 5-10 are best for growing this plant, but it can thrive in many conditions once it is established. In fact, you may have to keep an eye on this plant and put it somewhere where it cannot encroach on other plants, because it can become invasive in the right conditions. It does not like wet soil, and would prefer soil on the drier side. Be sure the soil you plant it in is well-draining. You can start seeds indoors and then plant them in your garden after the last frost. Try not to overwater these plants and plant them in an area with full to partial sun. Clip the aerial portions of the plants when they are in full flower.

If you are looking for a remedy for nerve pain, wilt the plants and then chop them (make sure you get as many flowers as possible in this remedy). Fill a glass jar with the plant material and cover it completely with olive oil, or a carrier oil of your choice. Let this infuse for four weeks, shaking it daily, before straining it out. Massage a liberal amount of this oil in areas where you are experiencing nerve pain and discomfort. If you want to use this plant for depression, follow the same steps as previously mentioned, but instead of using oil, use 80 proof or higher alcohol. Take 2.5 ml three to four times daily for management of seasonal affective disorder or mild depression.

8. Medicinal Plants to Have at Hand for Your Renal System

It is important to take care of your kidneys and drink plenty of water. This becomes even more important in a SHTF situation, because you certainly don't want to risk becoming dehydrated or giving yourself a kidney/bladder/urinary tract infection. Below are common plants you can employ to take care of kidney ailments and nourish the renal system.

a. **Dandelion (*Taraxacum of�icinale*) for cleansing**

This may be one of the easiest plants to grow in your garden because it will often "volunteer" in your garden without you lifting a finger! Most people think dandelion is a weed, but it is quite the opposite. Firstly, all parts of the plant are incredibly useful. You can eat the flowers and leaves for optimum nutrition. Second, the roots have potent cleansing properties and can act as a tonic for the kidneys. One simple way to get this plant to grow in your garden is to find the dandelion seed heads and scatter them where you would like to see them grow. You could also just harvest them as you find them in your yard, since they grow in yards across the world! Just make sure you don't spray your yard with toxic substances so you can have healthy yard "weeds."

Harvest the root, which is a somewhat large taproot. You will need a good spade to do this. Once you have collected a few roots, wash them off good and chop them up. Lay the chopped roots on a towel to dry. When they are sufficiently dry, bag them up and use them as needed in tea. Simply fill a reusable tea bag or tea ball with the root pieces and infuse it in a cup of hot water for several minutes. Drink one to three cups a day if you are in need of renal cleansing. Make sure to drink plenty of water, as the roots act as a diuretic and pull water out of the body to aid in cleansing.

b. Celery root for nourishment and cleansing

Celery is a common garden plant because it can provide a healthy snack, but it also has medicinal properties. Celery roots are diuretic and can help to eliminate waste from the body, especially the renal system! To grow celery from seeds, soak the seeds overnight. Next, plant the seeds in little trays of soil and place them in a window sill or sunny area of your home. A great idea is to cove the trays with plastic wrap after watering to trap in moisture and warmth. Remove the wrap when the seedlings begin to sprout. Keep the area moist and make sure they are getting sun as much as possible. Transplant them in your garden when the seedlings are stronger and the ground is warmer (at least 50 degrees Fahrenheit). Plant the seedlings at least eight inches apart and keep them well-watered.

Harvest the mature plant, root and all, for medicinal purposes. Don't waste the above ground parts – eat them! To prepare the root, chop it up good. Next, put the chopped root in a jar and cover everything with at least 80 proof alcohol. Let this infuse for four to six weeks before straining out and bottling the liquid. Take 2.5 ml up to three times daily for kidney cleansing. Drink more water than usual to counteract any chance of dehydration from the diuretic effects of the plant.

9. Medicinal Plants to Have at Hand for Your Reproductive System

Ailments of the reproductive system can be extensive, so it pays to pay attention to your body and what it is telling you. If you notice anything "off" don't take a chance and ignore it. Symptoms of hormonal issues in both women and men include acne on the jawline, mood swings, and pain in the reproductive regions. There are plants you can have on hand to nourish your reproductive system and balance hormones, should you feel they are off kilter.

a. Vitex (*Vitex agnus-castus*) for balancing female hormones

Vitex, or Chaste Tree, has already been mentioned in the "endocrine system" section of this chapter. However, since it is an adaptogen, it is especially handy for balancing hormones. It is popular among many women for its ability to restore balance to whatever hormones are off in their body. As a result, women who take Vitex often discover that their cycles return to normal (if they were abnormal) and their mood swings begin to dissolve. Follow the growing and medicinal preparation instructions in the endocrine system section. Take 2.5 to 5 ml of Vitex tincture up to three times daily, depending on the severity of the hormonal issues. You can slowly start to take less as you notice improvement. Give this medicinal preparation several months to work, because although it works well, the results may be slow in coming initially.

b. Ashwagandha (*Withania somnifera*) for male hormones

Ashwagandha is another adaptogenic plant already mentioned in the endocrine system section of this chapter for its ability to support the thyroid. In addition to what it can do for the thyroid, it is also a great supplement for men who wish to boost their stamina, energy, and balance hormones. Growing and medicinal preparation instructions can be found in the endocrine system section. Take 2.5-5 ml of ashwagandha tincture twice daily for any issues afflicting the male reproductive system. When combined with a healthy diet and plenty of exercise, ashwagandha can also significantly reduce sluggishness and low sex drive.

10. Medicinal Plants to Have at Hand for Your Respiratory System

Having a bad cough or dealing with a lot of drainage are never fun, but when you are experiencing a SHTF situation, this can be quite worrisome. There is always the chance that if it is not taken care of soon it may develop into pneumonia. Staying on top of drainage and mucous is important, so make sure at the first sign of these things you are ready with the following plants from your garden:

a. Oregano (*Origanum vulgare*) for preventing infections and clearing mucous

Oregano may be one of the most important herbs you can grow in your medical garden, and thank goodness it will also be one of the easiest to grow! It has long been hailed for its bacteria-killing abilities. It also contains constituents that help to loosen up mucous and get it out of the body. Oregano doesn't need much water to grow thick and healthy. It also isn't picky about the soil; just make sure to avoid really wet soil. Plant it in full sun for best results. Sow seeds in an area of your garden that will have plenty of room for this fast-spreading plant. Seeds should be sown six weeks before the last frost. Keep them watered, but do not overwater them. It can grow in nearly all USDA zones, but in zones that are prone to very hot, harsh summers, you may think about allowing the plant some afternoon shade to take the edge off. In no time you will have a garden full of this wonderful plant!

Clip the aerial parts of the plant to harvest for medicinal purposes. Hang small bundles to dry or place them on a drying rack. Store your dried oregano in a glass jar and keep it in a cool, dark place in your house. To use for respiratory issues, one easy and effective way is to create a steam using this herb. Boil a small pot of water on the stove and add a few tablespoons of dried oregano. Let this boil for a few minutes and then put your head over the steam (not too close, you don't want to get burned). Take deep breaths and inhale the steam coming from the pot. Put a towel over your head to help trap in more steam as you finish this breathing treatment. Make sure the towel does not come into contact with the hot burner or pot. Breathe in this steam for 10-20 minutes and up to twice daily for best results and to prevent any respiratory infection from forming. Keep doing this until symptoms are gone.

b. Horehound (*Marrubium vulgare*) for wet coughs and mucous clearing

Years ago, candy made from this popular herb was a favorite of boys and girls all over North America. While horehound may be known for its interesting flavor in candies, it is also a very handy herb for clearing mucous and calming a wet cough. It grows wild in many areas of North America and is very easy to grow in a garden! Use horehound for ailments like bronchitis, chest congestion, asthma, and allergies. Sow seeds in your garden around three weeks before the last frost. Cover the area with a dusting of soil to prevent the wind blowing the seeds away. Horehound does best in dry, poor soil. Do not over water this plant. It will not take long for horehound to take off once it is established. Be sure you keep an eye on it so it doesn't crowd other plants in your garden.

Harvest the aerial parts of the plant when you want to make medicine. You can dry some of what you harvest to make tea, or you can fill a jar with the plant, cover it in alcohol, and let this infuse for four to six weeks before straining out the liquid and bottling it. Take horehound in tea or in the tincture described at the first signs of congestion for best results. Drink two to three cups of the tea daily, or take 5 ml of the tincture up to twice daily for congestion. Do oregano steam treatments described above in addition to the horehound for expedited healing.

11. Medicinal Plants to Have at Hand for Your Skeletal System

Some common skeletal issues one may experience include sprains, fractures, and joint pain. Sprains and fractures can be very problematic in a SHTF situation, so you want to make sure you do not use the affected area and give it plenty time to heal.

You need your body in top shape to survive. There are several useful plants you can have ready in your garden to heal these ailments:

a. **Comfrey (*Symphytum of�icinale*) for sprains and fractures**

Comfrey is a great plant to have on hand in first-aid situations. This is because the leaves can be plucked and applied straight to a sprain, helping to relieve inflammation and pain in the immediate area. It was a staple in gardens for centuries for this reason. It is harder to grow from seed, so many gardeners use root pieces (comfrey grows easily even when small root pieces are planted in the ground). However, it may be hard to find root pieces, so if you wish to plant from seeds, you will need to stratify them first. To do this, place the seeds in a moist paper towel. Place this into a glass jar and close the lid. Place this in a cold place, like a refrigerator, for up to sixty days. Now your seeds should be ready to plant. Start them indoors if the temperatures outdoors are under sixty degrees Fahrenheit. Start them in well-draining soil and keep the soil moist but not soaked. Place your seeds in a sunny area of your house and let them get strong. When the weather is warm, place them in a sunny area of your garden. Plant them in an area where they won't encroach on your other plants, as they can become very invasive. They will come back every year.

To use them, pluck a leaf and apply to the affected area. It helps to wrap the area so the leaf stays on. Keep it wrapped for several hours and reapply leaves as needed until the swelling goes down. Stay off the area until further assessment determines what should be done.

b. **Cabbage (*Brassica oleracea*) for inflammation and pain**

This is another amazing plant that is usually thought of as food, and rightfully so!

Cabbage is nutritious and even more nutritious when it is fermented and full of beneficial probiotics. However, this unassuming plant also boasts impressive anti-inflammatory properties. Controlling major inflammation means controlling pain, as the two go hand-in-hand. Cabbage has long been used by nursing mothers who wish to wean without pain.

The same concept can be used with sprains, fractures, and sore joints. To grow cabbage, plant seeds indoors around six weeks before the last frost. Place them two inches down and keep the area moist and in full sun. Get your seedlings used to the cold by sitting them outside when they are established (before planting them outdoors). After the last frost, place seedlings in your garden 12 to 24 inches apart.

To utilize this plant, simply break off a large leaf and apply it to the sore joint or sprain. Wrap this area with plastic wrap or another material to keep it in place. Leave it on for up to three hours before repeating with a fresh leaf. When placed on engorged and painful breasts, cabbage leaves help bring down inflammation and dry up the milk fast.

12. Medicinal Plants to Have at Hand for Infections

One of the most dangerous medical situations in a SHTF situation will likely be an infection. Antibiotics will likely not be available. You will have to rely on highly antimicrobial plants to prevent and treat any issues that arise from bacteria invading the body and causing issues.

a. **Garlic for UTI, yeast infection, MRSA, and more**

Garlic has already been mentioned in the "circulatory system" section of this chapter, so make sure you check out how to grow it there. It has many uses, and treating infections is at the top of the list. Garlic has been shown to treat many infections, ranging from deadly to common. If you are injured and wish to avoid infection in the area, you can apply poultice garlic cloves to the area daily to keep bacteria at bay. For internal infections, you can dehydrate and powder garlic and place it in capsules to take. Taking five to six capsules a day may be just what is needed to clear up a UTI. For yeast infections, place the garlic clove (after removing the outer paper) directly in the vagina for several hours daily until the infection clears.

b. Oregano (*Origanum vulgare*) to fight respiratory infections, staph infections, and parasites

Oregano and garlic can both be used to combat infections, and using them together just speeds up healing even more. Growing and using oregano has already been mentioned in the "respiratory section" of this chapter, but there are other ways to use it for infections in other areas of the body. For UTI's and internal infections, you can chop garlic and place it in a glass jar, covering it in at least 80 proof alcohol to infuse for four to six weeks. Strain it out when the time is up and bottle your oregano tincture. Take 3-5 ml (depending on what your stomach can handle) every three to four hours until the infection is gone. It can be infused the same way as previously mentioned but in olive oil for external infections like wounds. Apply the oregano-infused oil to wounds daily. Another great use for this oil infusion is ear infections. Do not put this inside the ear, but rather massage around the outer ear and lymph area down the neck. This will help drainage in the ear, as well as treat infection.

*Another great infection-fighting substance you should consider having on hand is raw honey. Start looking into beekeeping now so you are prepared with plenty of honey if and when the time comes. Honey is highly antimicrobial and can be smeared on wounds to prevent and treat a variety of infections.

13. Medicinal Plants to Have at Hand for Pain Management

If you are one of those people who find yourself grabbing for the Tylenol or Ibuprofen, you will want to make sure you have alternative sources of pain management in your medicinal garden. These pain-relieving plants also have the bonus of less liver and kidney-ruining side effects, unlike the drugs previously mentioned.

a. Toothache plant (*Acmella oleracea*) for numbing and general anesthetic

You might be shocked at the numbing sensation that occurs when you place even the smallest toothache plant bud in your mouth and chew it a bit. The numbing is so profound that you may even find yourself drooling!

The plant comes by its name honestly and is a must-have for pain management. It is generally easy to grow, and an added bonus is that the leaves are edible and tasty in salads! To grow toothache plant, start seeds indoors or in a greenhouse in early spring. Make sure they are kept in a warm, sunny environment as often as possible. Keep the soil moist and it shouldn't take long for them to germinate. You can transplant them in your garden after the last frost, but be aware that they are not cold tolerant at all. Once cold weather hits, your plants will likely die. They will start producing lots of buds in late summer, so harvest buds daily when you are working in your garden.

The sphere-shaped buds are the medicinal portion of the plant. Set the buds on a towel to dry as you continue to collect them. Once you have a fair amount of buds, fill a glass jar and then top them with 80 proof alcohol, completely covering the plant material. Store it in a cool, dark place and shake daily. Strain out your new numbing solution and bottle it. Using a dropper, place 5-10 drops as needed in your mouth on any areas where you have a toothache or sore. It may even numb other areas of the body, depending on how strong your tincture is!

b. Valerian (*Valeriana of�icinalis*) for pain relief

Valerian may be famous for its ability to get you to sleep, but it is also an effective pain reliever. This is because it acts directly on the central nervous system. An added bonus is that valerian can also help to provide a sense of relief and calm to an unsettled mind. Instructions for how to grow valerian, as well as how to create valerian tincture, can be found in the "Nervous System" section of this chapter. For pain management, take 5 ml of valerian tincture every three to four hours as needed. Avoid operating any heavy machinery or partaking in any dangerous activity after taking valerian, as it may impair your ability to move quickly.

Many of the medicinal plant preparations in this section entailed creating an alcohol extract called a tincture. This is because tinctures have a very long shelf-life. Many tinctures will last up to seven years, and even as long as ten. This means that if you are able to grow many of the plants mentioned in this book in your garden and have a good harvest, you may be able to create enough tinctures to last you for years just in case your garden doesn't do as well later on! The alcohol acts as an effective preservative, as well as an optimal solution for extracting the medicinal portions of the plants. Of course, you can always harvest and dry your plants, making sure you store them in a cool, dark place in airtight containers afterwards. Then they will be on hand when you need them for teas and poultices. However, keep in mind that dried herbs have a shelf life of around one year.

It might be a good idea to start looking into moonshine stills and learning how to make your own alcohol for tincturing purposes.

Estimated Cost

Costs for creating your own medicinal plant garden are going to vary greatly depending on many factors. Lumber and soil prices will be different in different states and may even vary depending on where you are purchasing them. If you are the do-it-yourself, self-sufficient type, you are going to save a lot of money in lumber (by cutting your own) and on soil (by using soil from your property). You will also save on construction by building the beds yourself and not paying to have them built.

If you wish to purchase lumber for your beds, understand that the price of lumber has gone up in the past year and you will likely be paying more than you normally would. You have several options when it comes to lumber for your beds. You could go for longevity and purchase plastic lumber that will hold up for years and not rot away. It would cost you anywhere from $200-$250 per bed if you go this route. The pro is that it is durable and will last a while. The con is that it is more expensive than using untreated lumber.

It will cost roughly $100 per bed if you use untreated lumber. A pro is the affordable price, but the con is that untreated lumber may not last too many years before it begins to rot. You can buy treated lumber of course, but it is highly unadvisable to use chemically treated lumber for a medicinal plant garden. Those chemicals can leach into the soil and do more harm than good with your plants and your health.

Aside from lumber, soil will be your next big expenditure. If you cannot get soil from another area of your property to add to your beds, you can order a dump truck of soil to be poured in the beds. Prices vary, but you should be able to order what is called a "tandem dump truck load" of topsoil delivered to your house for around $45 to $60 per raised bed. This is roughly 10,000 cubic feet of topsoil. This option is absurdly cheaper than going out and buying bags of topsoil and trying to fill your beds. If you were to attempt this, you would be spending hundreds of dollars more than just getting topsoil delivered.

For essential tools and pest control (mentioned earlier in this chapter), you can safely budget $100 to $200 to get what you need to dig up the beds, plant seedlings, control pests, and weed your garden.

Here are some useful links to where you can purchase the needed materials from:

- Pruning Shears:
www.amazon.com/Fiskars-91095935J-Bypass-Pruning-Shears/dp/B00002N66H/

- Trowel:
www.amazon.com/Mr-Garden-Trowel-Stainless-Gardening/dp/B08FHN6WSP/

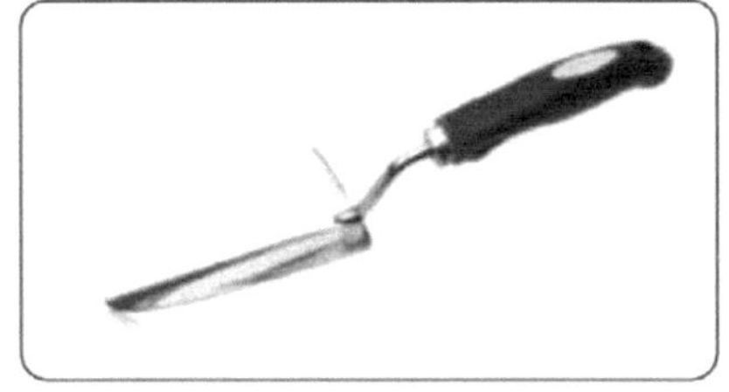

- Weeder:
www.amazon.com/Grampas-Weeder-CW-01-Original-Remover/dp/B001D1FFZA/

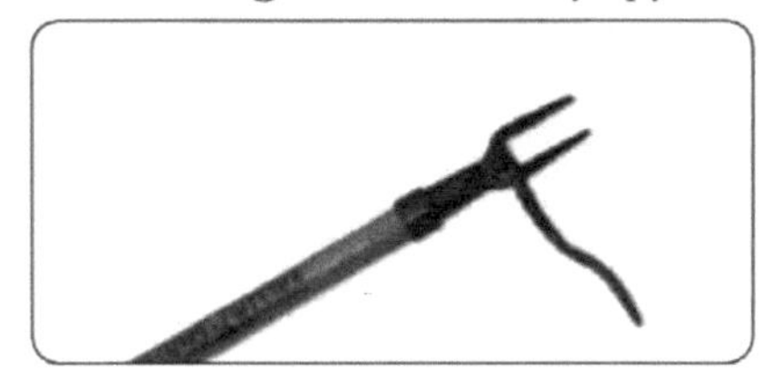

◆ Garden Tool Set:
www.amazon.com/Aluminum-Lightweight-Gardening-Anti-Skid-Ergonomic/dp/B08C7HXJR9/

◆ Rake:
www.amazon.com/Martha-Stewart-MTS-TELR-Comfort-Handle/dp/B086FG3XR7/

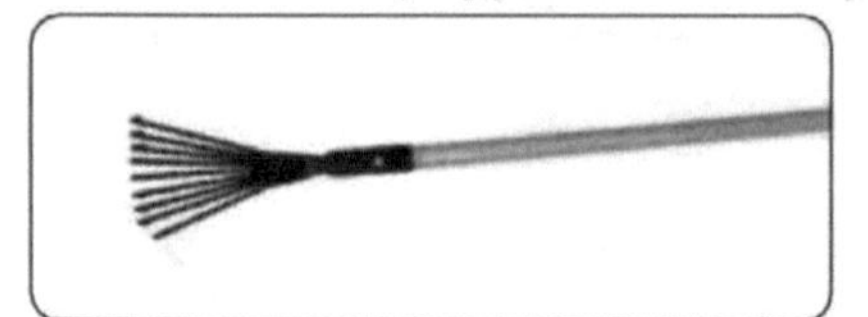

◆ Hoe:
www.amazon.com/Berry-Bird-Stainless-Weeding-Cultivating/dp/B08C4KTSS5/

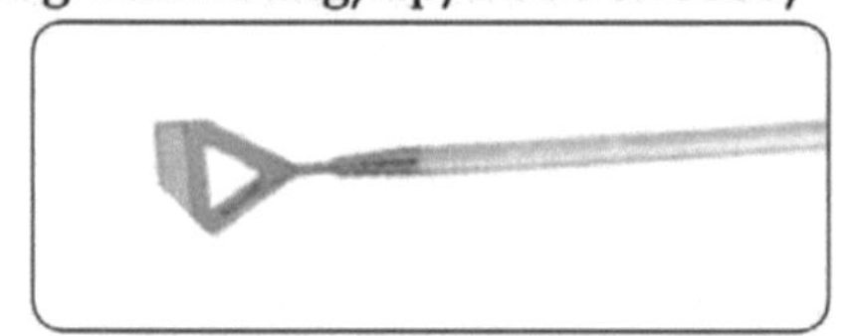

◆ Moisture Meter:
www.amazon.com/SURENSHY-Gardening-Hygrometer-Humidity-Required/dp/B0895QM8S9/

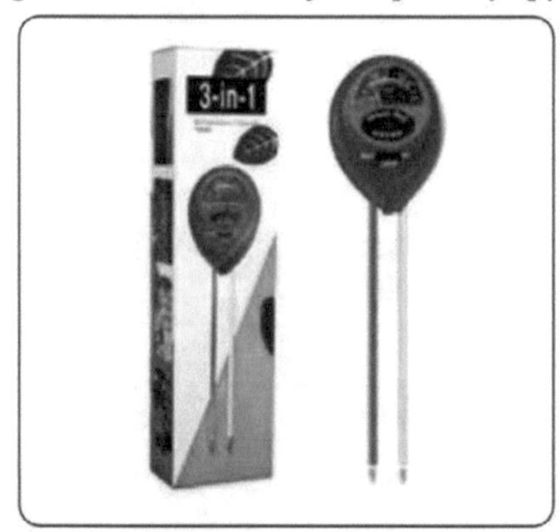

◆ Seed Dispenser Kit:
www.amazon.com/Batino-Adjustable-Transplanter-Spreaders-Dispenser/dp/B07X23YRXS/

Seed Trays:
www.amazon.com/9GreenBox-Seedling-Starter-6-Cells-Labels/dp/B0149L72CE/

◆ Diatomaceous Earth for Pest Control:
www.amazon.com/Harris-Diatoma-ceous-Earth-Food-Grade/dp/B07RV67ZNL/

◆ Neem Oil for Pest Control:
www.amazon.com/Harris-Pressed-Unrefined-Cosmetic-Grade/dp/B07732SVD3/

◆ Lumber:
www.homedepot.com/b/Lumber-Composites/N-5yc1vZbqpg?catStyle=ShowProducts&NCNI-5&search

*Top soil will be locally sourced, so you will need to contact a company in your area that delivers topsoil by the truckload.

Seeds may be the third largest expenditure, after lumber and soil. You want to make sure you are purchasing non-GMO, preferably heirloom seeds from a reputable seed company that specializes in medicinal seeds. There are several great companies that you can purchase seeds from. You may also choose to give your business to your local nursery. Prices vary, but plan on spending anywhere from $150 to $250 if you want to purchase all the seeds mentioned in this chapter. This may seem pricey, but keep in mind that you can easily save seeds from the plants you grow the first year so you don't have to keep buying them every year thereafter.

How Much Work to Put Into it

Your garden can take a lot of work to get started, but keep in mind that the more work you put in, the less money you are going to spend creating it. If you can put in the work to chop down your own cedars, build your own beds, and get your own topsoil, you are cutting out hundreds and hundreds of dollars. If you put your mind to the task and have the whole day to work on it, it will probably take you no more than a week to get things built and ready to go.

You will not be able to avoid the work that goes into starting seedlings indoors or in a greenhouse. This requires time and patience. It will also take time transplanting seedlings into your raised beds. If you have back issues, take your time with this part because it will require you to be hunched over planting for a while. Rest in between plants and walk a few circles around the bed before getting back to it. If you have bad knees, consider getting a mat to kneel on as you plant.

Reserve time to visit your garden daily to keep up with the weeding and pest control. It will not take long for weeds to take over an unattended garden. If you can, purchase a weeder so you don't have to bend down and pick weeds all the time. Make sure you keep your plants watered, but many of the plants mentioned in this chapter do not need watered daily unless it is extremely hot and dry in your local climate. Going a day or two between watering can actually help the roots grow stronger and deeper as they search for water.

How to Tend Your Garden in Winter and How to Use Plants in Winter

If you plant the plants mentioned in this chapter, many of them will die back in the winter but come back in the spring. Make sure you take notes on which plants will come back and which will not so you know what to plant the next year and what not to plant.

When the weather begins to cool down before the first frost of the year, go out and harvest all of your remaining plants before the frost kills them. Remember to check back to the plants section of this chapter to know which parts to harvest and which to leave. Some plants will only need to be trimmed down, while other plants will need dug up because the root is the medicinal portion of the plant. Also keep in mind that a few of the plants mentioned in this chapter require more than one year's growth before harvesting. In that case, leave them be until they come back next year.

Let some of your plants go to seed. Harvest those seeds to plant next year. This will ensure that you will always have medicinal plants to use on your homestead, as well as save you a lot of money on seeds. Set your saved seeds on a towel to dry before storing them in a container in a cool, dark place until they are ready to plant.

The majority of your harvested plants do not need to be processed immediately, although if you are tincturing them in alcohol, they can be processed right away if you want. If you don't have a lot of time on your hands to tincture a bunch of plants that you just harvested, you can hang them somewhere to dry, and then when they are fully dry store them in an airtight container in a cool, dark place until you are ready to create a tincture or make tea. If at all possible, try to tincture all of the plants that need tinctured (see plant section of this chapter for instructions) because this will extend the shelf life and give you medicine on hand to use as soon as you need it. Remember that tinctures and oil infusions take time to make.

As spring comes back around, keep an eye on emerging plants coming back after the long winter. If you start to notice any plants sprouting and are expecting another frost soon, cover them with a tarp to keep them from being killed. With a little attention and love, you will likely find your garden rewarding in more ways than you imagined.

How To Stockpile Seeds

During the height of the Covid-19 pandemic, gardeners were in for quite a shock as they walked into some retail stores and found the seed aisle blocked off. Many big retail chains that remained open during the pandemic seemingly decided that seeds weren't important enough to allow the public access to buy. In fact, it's one year later and some retailers are still not allowing access to seeds. Is it really because they are deemed "non-essential"? Could there be other motives behind this insanity?

The sight of caution tape strung all over the seed aisle made a lot of free-thinkers stop and think. Why deny access to seeds? Have we forgotten where our food comes from? Have we reached a point in society where one cannot grow their own food, but instead must rely on grocery stores to supply what we need? This is a dangerous precedent.

Things were not always this way during times of crisis. During times of war, specifically WWI and WWII, the government actually encouraged the public to grow a garden. They were termed "Victory Gardens" and were planted at both private residences, as well as public parks throughout the United States, United Kingdom, Australia, Canada, and Germany. These gardens were encouraged because of a serious food shortage. Those who worked in the agricultural field were called into the service to fight. This put a strain on manpower to maintain enough crops to feed the country.

All land not being used was suddenly worked and turned into gardens. Self-sufficiency was highly encouraged with the slogan "Grow your Own, Can your Own". A victory garden was even planted on the White House lawn. This attitude of self-sufficiency and growing your own food to care for your family has faded away over the years to the point where the current generation has little to no knowledge or skill when it comes to seeds, gardening, and how to survive in a crisis.

There may come a time when grocery stores don't have what you need to survive. Working now to be as self-sufficient as possible will pay off exponentially in the future. Seeds are your ticket to provision for yourself and your loved ones.

Why You Should Stockpile Seeds Now

Seeds can and should be stockpiled now more than ever. When the powers that be are telling the public not to buy seeds, that is when you should buy as many seeds as you can. Start your collection now and keep collecting at every opportunity.

One of the biggest incentives to start stockpiling seeds is what has come to light during our recent time of crisis. A pandemic hit the world in 2020 and we had the opportunity to see how people reacted to this time of turmoil. What did most people do? They stockpiled things like toilet paper. Toilet paper is something that is nice to have, but we can see where people's priorities were by the empty shelves in the toilet paper aisle. What about basic necessities like food and water? Those were also stockpiled in many cities around the world. Even today, the shelves in some produce and meat aisles are surprisingly scant.

This mass drive to buy things that are not necessary to our survival goes to show that the public can react fast and irrationally in a crisis. While people are out buying all the bread and milk, you can take comfort in the fact that you have access to a bounty of nutritious food in your backyard for years to come, all without ever leaving your home. Most food items at grocery stores have a shelf life. By the time you purchase the food and bring it home, you have a few weeks at best. When you have your own survival garden with the seeds you collect, you have fresh food year-round just a few steps away. The time to stockpile seeds has certainly come and the importance of being prepared with plenty of seeds cannot be ignored as we face an uncertain future.

The Best Way to Stockpile Your Seeds

Believe it or not, most seeds have a decent shelf life. Their survival all depends on the conditions in which you choose to store them. You can very realistically collect enough seeds to last your whole life.

Temperature and humidity are everything. Make sure that wherever you choose to keep your seeds, the humidity is low to non-existent. The temperature needs to be cool to room temperature. High humidity can negatively impact seeds because it increases the chance they are exposed to some moisture. Seeds should not be exposed to any moisture until they are ready to be planted. Temperatures too hot or too cold can damage seeds as well, although some seeds don't mind.

Make sure you store your seeds in a container that is completely water proof. Water is their enemy until they are ready to be planted. Another thing to keep in mind is that the container should be as airtight as possible. Most seeds come in individual envelopes. This is completely fine and there is no reason to remove them from these if they are dry. These envelopes can be stored in an airtight container with a lid for safe keeping. Containers like waterproof five gallon buckets with securely-fitting lids are perfect.

Do you ever notice little silica gel packets in products you buy? Save them! These packets are placed in products to keep moisture from ruining them. You can place some of these packets inside your containers to keep moisture at bay. They can also be purchased online.

Store your seeds in a place with consistent, non-fluctuating temperatures. Make sure the area is dry and dark. Sometimes, sunlight can damage products and it may also heat up the container and damage seeds.

Label your containers. This is very important. This saves you from having to constantly open the containers and rifle through seeds to try and find what you need. Seeing what you have will also allow you to take stock of what you have and what you still need to collect. Write the date you purchased the seeds, as well as the type of seed, on the outside of the container. This will help you keep track of which seeds need to be used first. Always plant the oldest seeds first.

Seeds can be stored in the refrigerator or freezer for safekeeping and to extend shelf life. These are best stored in envelopes inside glass mason jars. Fill a jar (write all information on the outside of the jar) and place it in the freezer until you are ready to plant. There is a process to removing these seeds and planting them. First, carefully sit the jar out at room temperature for at least twelve hours. Remove the lid and let the air circulate in the jar so it can prevent any moisture that may occur from the temperature change. Let this jar sit in the open air several days before planting the seeds.

Seed Shelf Life

There are some seeds that last longer than others. Seeds that last up to two years before their germination rate decreases include: sweet corn, onions, parsley, peppers, parsnips, and okra. Keep an eye on the dates you purchased these particular seeds and try to use them in a timely fashion.

For all seeds you purchase, try to buy heirloom seeds. These are the best seeds for growing food that you can then collect more seeds from to save. For the seeds with a shorter shelf life, like the ones mentioned above, plant them each year and try to have enough for one or two more years in your collection. Each year, collect seeds from the vegetables you grow and save them by harvesting them, cleaning them, and drying them thoroughly. When they are completely dry, place them in labeled envelopes and store them in a container. Always plant your oldest seeds first.

Seeds with a shelf life of up to four years include: carrots, celery, peas, leeks, eggplants, broccoli, kohlrabi, cauliflower, Brussels sprouts, beets, beans, watermelons, pumpkins, turnips, spinach, tomatoes, turnips, and squash. Try to have around four years' worth of these seeds on hand and use the oldest seeds each year during planting time. It is not a bad idea to collect these heirloom seeds each year to process and store for emergencies in case something happens to your collection.

Some seeds have a shelf life up to six years, or even longer, when stored properly. These include cucumber, radish, and lettuce seeds. Try to always have up to six years' worth of these seeds on hand in your stockpile.

There may come a time when you are no longer able to purchase seeds. If this happens, you should already be prepared with all the heirloom seeds you have grown and processed for storage. Each year, whether you have ample purchased seeds in storage or not, collect seeds from your garden to dry and store just in case. If you can keep doing this through the years, you will start a self-sufficiency cycle that can last a lifetime.

The Best Seeds to Stockpile

The best seeds to stockpile in terms of shelf life include cucumber, radish, and lettuce seeds because they last longer than most other seeds. Under the right conditions, they can last much longer than the six year shelf-life they are generally given.

In terms of optimal nutrition, the best seeds to stockpile include amaranth, spinach, alfalfa, kale, broccoli, peas, and beets. These are packed full of vitamins and minerals your body needs to flourish. They may come in especially handy when you can no longer purchase vitamins to take daily.

In addition, consider making sure you have an ample amount of starches, such as sweet potatoes, potatoes, carrots, pumpkin, squash, and corn. These are not only nutritious, but they contain fiber that can help keep your digestive system healthy.

Always have seeds on hand that produce vegetables that supply ample protein to the body. Protein is important for the body to function. It also helps the body maintain healthy muscle mass. Some healthy sources of protein include chickpeas, lentils, asparagus, pinto beans, lima beans, and fava beans. Quinoa, chia seeds, and Brussels sprouts also contain protein.

Stockpiling Seeds for a Brighter Future

While the future is uncertain, don't let fears or anxiety about the future steal your joy. While things may look bleak at times, it is within us as human beings to forge on and blaze a path to victory. Stockpiling seeds give you peace of mind, optimal nutrition, and the ability to survive most any crisis without depending on help from the outside world. This independence and self-sufficiency is one of the most precious gifts you can give yourself and your family.

Natural Remedies To Make At Home Using Local Plants

Did you know that there are numerous plants growing near you that can be used to make natural remedies for all kinds of ailments? The plants in this chapter are familiar and some are even invasive. They can often be found growing right outside your door in many parts of North America. Below, you will discover what plants you can use, as well as how to use them for common ailments.

1. Natural Remedies to Fight Viruses

Japanese Honeysuckle, *Lonicera japonica*

Japanese honeysuckle is a vining plant that has become an aggressive and invasive plant in many parts of the world. However, this "weed" has a surprising medicinal use for killing viruses. All aerial parts of the plant can be utilized to create a tincture to target a variety of viruses. Tinctures are plant extracts that use a liquid solvent like alcohol to pull the medicinal properties out of plants.

To create a tincture with this plant, rip it out (vine and all) and chop what you have collected into small pieces. Make sure to harvest the plant when it is flowering. Fill a glass jar with the vine, flower, and leaf pieces and then cover the plant material completely with at least 80 proof alcohol. Let this sit in a cool, dark place for four to six weeks. Place a lid on the jar and shake it daily to help the tincture infuse better. After four to six weeks, strain out the liquid and bottle it. Since they are made with alcohol, tinctures have a long shelf life and can last for up to seven years if stored out of sunlight and in a cool place. Take 5ml every few hours as needed if you feel you have a virus.

Elderberry, *Sambucus nigra* or *Sambucus canadensis*

Elderberries are a powerful antiviral. So many people are unaware of just how common elder is in North America. It flourishes in the wild all around us. It can easily be spotted in midsummer because the big umbels where the berries will emerge are clusters of white flowers. Look for bushes with these large white flower clusters and take note of where you saw them. Go back to those spots in early fall and check for berries. The berries are ripe when they are a dark purple color. Collect the berries by trimming the stems and then lay the stems in a dehydrator to dry the berries. Once they are dried, they will fall off the stems easily and will be much easier to prepare. Fill a glass jar with the dried berries and then cover them completely in 80 proof (or higher) alcohol. Let this infuse for four to six weeks, shaking the tincture daily to help it infuse. It will eventually turn a deep crimson or purple color. Strain it out when it is ready and bottle the liquid. Take 5ml two to three times a day at the onset of a virus. The sooner you begin taking it the speedier your recovery will be.

2. Natural Remedies to Treat Wounds

Yarrow, *Achillea millefolium*

Yarrow is commonly found in early to late summer. It is usually found growing in fields and pastures. It can be identified by its white flower cluster on top. It is an uneven shape, unlike Queen Anne's lace, which is more round. Another characteristic that makes yarrow easy to identify is its medicinal and herbaceous aroma, unlike Queen Anne's lace which as a carrot-like aroma.

Yarrow is known for its ability to heal wounds; specifically, wounds that are bleeding. It can act as a styptic and help the blood to clot. Additionally, yarrow can help cleanse a wound with its antiseptic properties. Yarrow can be used in several ways, but one popular way is straight from the plant to the wound. Pluck a leaf and mash it up. Then place it on the cut. You can also collect yarrow and dry it for later use. Once it is dry, grind it into powder and apply the powder to wounds.

Witch Hazel, *Hamamelis vernalis*

Witch hazel is native to many parts of North America. This tree may not be easy to spot during the spring, summer, and fall months. However, it is easy to spot in the winter due to its bright yellow blooms that emerge when few other colors are visible in the forest! Take a walk in the woods on a winter's day and look out for witch hazel trees with their yellow, ribbon-like flowers. You may even notice a pleasant aroma before you get to the tree itself.

This tree has been utilized for ages for its astringent, wound cleansing properties. It is also known for reducing inflammation and redness on the skin. It can help calm angry skin and promote healing. Harvest a few small branches and let them dry in your apothecary. When you need them, you can break off some to infuse in hot water to make tea. This tea, when cooled, is useful for wound cleansing. This tea is very soothing when applied to hemorrhoids and postpartum wounds as well.

3. Natural Remedies to Treat Allergies

Purple Deadnettle, *Lamium purpureum*

One of the first plants to emerge in the spring, besides dandelions, is purple deadnettle. It is a member of the mint family, but does not have a minty smell. It can take over a yard from March to May. It has tiny purple flowers emerging from the top of a square-stalked plant. It does not grow too big, maybe up to five inches in height. The leaves are toothed and pointed at the top. They may have a purple color the closer to the top they are.

This plant contains flavonoids that can help suppress histamine production in the body. Collect plants from an unsprayed yard and chop them well. Fill a glass jar with the plant parts and then cover them completely with alcohol or apple cider vinegar. Let this infuse for four to six weeks before straining out. Take 5ml as needed to help with seasonal allergies.

Stinging Nettle, *Urtica dioica*

Stinging nettle also contains flavonoids that help to suppress histamine production, thus lessening allergy symptoms. It can be found in the summer and fall months. Harvesting these plants means wearing gloves to protect your hands from the little stingers these plants possess. If these stinging hairs make contact with your skin, they can cause a lot of irritation. They can usually be found growing in fields and waste areas. They can become aggressive in many areas, so they usually aren't hard to find.

Dry the nettles by hanging them in an area of your house and when they are sufficiently dry fill a tea infusion ball or tea bag with crumbled leaves. Let this infuse in a cup of hot water for several minutes and then enjoy. Drink a cup of stinging nettle tea as needed for allergies.

4. Natural Remedies to Treat Sore Throat

Echinacea, *Echinacea* spp.

This native pink flower is a powerful remedy against a sore throat caused by a bacterial infection. There are several species of Echinacea that grow throughout North America, with *E. purpurea* being the most popular for medicinal use. *E. angustifolia* has also been used for similar purposes. This plant is easily identifiable in the summer because it has bright pink flowers with a spiny black center. The plants are often on the taller side, ranging from three to five foot tall. When harvesting in the wild, be mindful and only take what you need. Do not harvest roots unless there are many flowers around.

Use all parts of this plant, from the flowers to the leaves to the roots. Chop everything well and fill a glass jar with the plant material. Next, cover it with at least 80 proof alcohol and let this infuse for four to six weeks. Strain everything out and bottle the liquid when it is ready. For strep, gargle for one minute in the back of your throat and then spit it out. Do this every other hour for a week until the strep is gone. You may also consume 5ml up to three times a day to help boost the body's immune system to help combat strep throat.

Bee Balm, *Monarda fistulosa*

This native plant contains a constituent called thymol, which is highly antibacterial. It is even used commercially in some sanitizing products to kill germs. Bee balm also goes by the name wild bergamot in many parts of North America. There are several species, with some being lavender in color and others a dark magenta. They can be identified in the summer months by their wild-looking flower heads that are comprised of many small, tubular stalks coming from a central head.

Use the aerial parts of this plant to create a tincture to use for killing bacteria that cause throat issues. It can be gargled or taken internally in 5 ml doses (up to three times daily). It is extremely effective against throat infections when used along with Echinacea.

5. Natural Remedies to Treat Ear Infections

Mullein, *Verbascum Thapsus*

It is a popular anti-inflammatory and pain-relieving remedy that can be found growing in the summer months and can be identified by its fuzzy, large leaves. A stalk comes from the center of the basal rosette of leaves. The yellow flowers can be collected from these stalks.

Once you have enough yellow flowers collected, fill a glass jar and then cover the flowers in olive oil. Let this infuse for four weeks. For extra antibiotic strength, you can add a dried and chopped garlic clove. When the time is up, strain out the oil and keep it refrigerated. When you have an ear ache or ear infection, take the bottle of oil and gently warm it by placing it in a bowl of hot water until the oil is melted and warm. Place a few drops into the affected ear and let this sit for several minutes. Do this as needed throughout the day to combat infection and inflammation.

Oregano, *Origanum vulgare*

Oregano-infused oil is another effective natural remedy for ear ailments. Oregano is a very common garden herb, but contains powerful medicinal properties. There is a wild native plant called American dittany, *Cunila origanoides*, that contains a very similar chemical profile to oregano and can be used the same way. It smells identical to oregano and has a woody stem, lance shaped leaves, and small purple flowers that emerge in the fall.

Collect the oregano or dittany and let dry by hanging it somewhere with good air flow. Next, chop the aerial parts well. Fill a jar and cover them in olive oil. Let this sit for a month before straining it out. Apply this infused oil to the area around the ear (not inside the ear!) and massaging the lobes, around the ear, and the lymph area down the neck from the ear. Do this multiple times a day to help with drainage, kill infection, and provide pain relief.

6. Natural Remedies to Treat Bronchitis and Respiratory Ailments

Mullein, *Verbascum Thapsus*

Mullein leaves also have a medicinal purpose. These thick and fuzzy leaves are an excellent remedy for mucous expulsion and to help clear bronchial passages. You can make a tincture from the leaves by collecting them from a clean area, chopping them, and covering them in alcohol in a glass jar. Let this infuse for four to six weeks before straining it out. The liquid should be a medium to dark green color. Take 5 ml every few hours to help your body get rid of mucous and clear the airways.

Pleurisy Root, *Asclepias tuberosa*

Another name for this familiar plant is "butterfly weed" or "chigger weed." It can be spotted in the summer months by its bright red, vibrant clusters of flowers. It is a great pollinator plant, so you will often notice butterflies surrounding it. It only stands a foot or two high and can be spotted in fields and pastures throughout North America.

The root of this plant has been used as a lung remedy for centuries. It can help to clear the airways, expel mucous, and calm irritated lungs. Tincture the roots by harvesting them mindfully, chopping them, and covering them in alcohol in a glass jar. Let this sit for four to six weeks before straining it out. Take two to five ml every three to four hours if you are suffering from a respiratory issue. You can use this in conjunction with mullein leaf for additional healing.

7. Natural Remedies to Treat Urinary Tract Infections

Oregano, *Origanum vulgare*

Because oregano contains potent antibacterial properties, it can be used to kill a variety of bacterial issues in the body. One such issue is a urinary tract infection. This results when bacteria get in the urinary tract and cause irritation. Drinking tea made from dried oregano leaves can flush out the bacteria, as well as kill it. You can infuse the leaves using a tea bag or tea infusion ball in a cup of hot water for five to seven minutes. Drink as many cups as you can throughout the day to target the infection and flush it out.

Aloe Vera

Aloe vera is native in many southwestern states, but those who don't have it growing wild will often have it growing in their house. Aloe vera is a very common house plant because it is so easy to grow and requires little maintenance. It spreads easily and can propagate from little "babies" that spring up around the plant.

If you have this plant, you can use it to your benefit by slicing open the leaves and scraping out the inner gel-like substance. Drink a tablespoon of this in a smoothie or a glass of water up to three times a day. This will help to relieve inflammation in the urinary tract, as well as kill the infection. Using this in conjunction with oregano tea is optimal.

8. Natural Remedies to Treat Gastrointestinal Issues

Mints, *Mentha* spp.

Not just one mint species is best for helping relieve gastrointestinal discomfort like gas, bloating, upset stomach, stomach cramps, etc. All members of the mint family can help. Some great examples to look for include include peppermint, *Mentha piperita*, catnip, *Nepeta cataria*, spearmint, *Mentha spicata*, and horsemint, *Mentha longifolia*. You can get a good idea whether or not you have a mint by looking at the stalk of the plant (which should be square) and by smelling the crushed leaves. Avoid germander, which is a member of the mint family that grows in the summer months and has light pink flowers atop stalks. It can be potentially dangerous. Harvest aerial parts of the mint plant and hang them to dry. When they are sufficiently dried, store your mint in a glass, airtight jar in a cool, dark place. When you need relief from gastrointestinal issues, simply take some dried leaves and infuse them in a cup of tea to drink as needed. You will find that this can help relieve bloating, gas, upset stomach, and constipation.

Agrimony, *Agrimonia eupatoria*

This common and native plant can help provide relief from diarrhea and promote gastrointestinal healing. Agrimony can be found at the wood's edge in the late summer months. It can be distinguished by its long stalks with small yellow flowers at the top of the spiked stalks.

Harvest the aerial parts and dry them for use when you need them. You can infuse the plant parts in a cup of hot water to drink for diarrhea. Drink as needed when you need relief.

9. Natural Remedies to Treat Pain

Willow, *Salix* spp.

Willow bark is known for its ability to combat pain and treat headaches, body aches, and inflammation. There are several species of willow that can be used because they contain salicin, a medicinal compound. The most popular species of willow used medicinally is the white willow tree, *Salix alba*. However, there are other species that can be found all over North America and be used medicinally.

The bark is what contains the compound. Cut into the inner bark of the tree to get what is behind the firs layer. Do not cut a ring completely around the tree or you could kill it. You can use this bark to infuse in water to drink as a tea for pain or make a tincture with alcohol for pain. For tea, drink a cup every few hours as needed. For a tincture with alcohol, take 5-8 ml every four to five hours for pain from headaches, body aches, or tooth aches.

Poplar, *Populus* spp.

Like willow, the bark from the poplar tree also contains salicin. This tree can be found all over the world, primarily because there aren't many locations it can't grow. Harvest the inner bark from this tree much the same way you would willow bark. It can be used in the same ways, whether in a tea or tincture for pain.

10. Natural Remedies to Treat Circulation/ Heart Ailments

Motherwort, *Leonurus cardiaca*

It can be found growing wild in many areas of North America, and is also a lovely garden plant. It is a member of the mint family and can be identified in the summer by its stalks with protruding, lobed leaves and purple flowers going up to the top. It has a pleasant, minty smell. It is known for its ability to nourish the heart and circulatory system. It can help lower blood pressure and strengthen the heart tissues. In addition, if blood pressure spikes are caused by stress, motherwort can help to provide calm and peace, thus lowering blood pressure as well.

Harvest aerial parts and chop them well. Cover the plant material in alcohol in a glass jar and let this infuse for four to six weeks before straining it out. Take 5 ml if you are feeling stressed or anxiety. This can help lower your blood pressure if you are worried it may spike as a result of the stress. Take 2.5 ml twice daily for heart and circulatory health.

Hawthorn, *Crataegus laevigata* and *Crataegus pruinose*

It can be found all over Europe and North America. The berries from this tree are used to help with heart and circulatory issues. Hawthorn is known to help lower blood pressure and nourish the heart muscle.

Harvest the berries anywhere from mid spring to early fall. You can tincture them and take 5 ml up to twice daily for heart and circulatory health and maintenance.

11. Natural Remedies to Treat Liver Issues

Dandelion, *Taraxacum officinale*

There may not be a more common, yet underrated, plant than dandelion. All parts of this plant can be used for both food and medicine. The root however, is known for its ability to tonify the liver and kidneys. You can find this plant almost anywhere, and it grows throughout the year. You may even be able to harvest it in the winter if the winter is mild. Use a good spade to dig up the plant, taking care to get the entire (sometimes large) taproot. Make sure you collect your dandelion roots from areas you know have not been sprayed with pesticides or other harmful chemicals. Wash the roots and chop them well. Let them dry and store them in an airtight jar for use in tea. Drink one to two cups of dandelion root tea daily for liver nourishment and healing.

Yellow Dock, *Rumex crispus*

This common plant flourishes in pastures and fields throughout the United States. It can be identified by its large leaves with curled edges and tops that go to seed in the summer and fall months, appearing dark red to dark brown in color. While the seeds are a fiber-rich and nutritious snack, the roots are the part used for liver health.

Dig up the roots in the fall and make sure you have a good spade because they can be very large and deep in the ground. When sliced open, the roots are yellow inside. They are thick and meaty. Wash the roots and dry them. Infuse the chopped and dried roots in hot water to make tea for liver health. Enjoy this daily to help flush the liver and keep it healthy.

12. Natural Remedies to Treat Kidney Issues

Cleavers, *Galium aparine*

They are common in the spring months.They are identified by their small, lobed leaves circling the stalk. One of the most common identification characteristics is the sticky nature of the plant. It will stick to almost anything, thus giving it the name "cleaver."

This plant is an excellent lymphatic cleanser, as well as an effective diuretic. Diuretic plants help to flush the body of toxins by promoting the removal of water from the body. If you wish to utilize cleavers for kidney flushing, simply harvest them and infuse them in alcohol while they are still fresh. Strain out the tincture after four to six weeks and take 5-10 ml of this up to three times in a day to cleanse the kidneys. Always drink extra water when you are taking a diuretic herb to avoid dehydration.

Goldenrod, *Solidago* spp.

It is common in the fall months and is characterized by its blazing golden flowers atop a stalk. It is a favorite for bees and other pollinators during the autumn months.

Goldenrod is excellent for inflammatory conditions in the kidneys because it can help to reduce inflammation and promote healing. In addition, goldenrod is a diuretic and can help flush out toxins. Gather the aerial parts in the fall and infuse them in at least 80 proof alcohol for four to six weeks before straining out the tincture. Take four to six ml of the tincture up to three times daily for kidney ailments like kidney infections or kidney stones. Drink plenty of extra water when taking this remedy to flush out the infection or stone. There are many species of goldenrod, but some popular medicinal species include *Solidago virgaurea*, *Solidago canadensis*, *Solidago gigantea*, and *Solidago odora*.

13. Natural Remedies to Treat Insomnia, Depression, and Anxiety

St. John's Wort, *Hypericum perforatum*

You may notice the bright yellow flowers of St. John's Wort in mid-June. It prefers sunny areas like open hillsides and pastures. You can even find it along roadsides. It grows in groups and can get up to three feet tall. It is a bushy native plant. You can tell it is the "real thing" if you smash one of the yellow flowers and your fingers become stained with a red-purple tint. This is proof the flowers contain hypericin, a chemical constituent in St. John's Wort responsible for its medicinal properties.

Harvest the flowering tops and tincture them in alcohol as soon as possible. The alcohol tincture will turn a bright red due to the hypericin. This plant has been shown to help relieve mild to moderate depression, as well as anxiety. Take 5 ml twice daily (morning and evening) to help manage these, as well as Seasonal Affective Disorder.

Skullcap, *Scutellaria lateriflora*

This is a beautiful wildflower that can be found in the summer months. It prefers well-drained fields and meadows. The flowers are a lovely blue-purple color and have a distinctive "hood." This may be one reason they were named "skullcap." They are a member of the mint family, so the stalk is square.

Harvest the aerial parts of this plant and tincture it in at least 80 proof alcohol for four to six weeks. Strain it out when the time is up. You should have a vivid green tincture. Take three to five ml of this as needed for anxiety and stress. Skullcap has been used for hundreds of years to help calm the body and mind. It acts on the nerves, helping to provide peace amidst frustration and agitation.

The Only 7 Seeds You Need To Stockpile For A Crisis

If you have ever visited the seed store, you may have noticed the sheer variety of seeds available to purchase. There are so many seeds available that it may seem overwhelming when trying to decide which seeds are the best to purchase for your needs. Don't stress! The research has been done for you. Below you will discover the seven best seeds you need to stockpile for a crisis. They have been chosen for their nutrition content, ease of growing, and versatility of use. You will learn about why these seeds are superior, what they can do for you, and how to grow them.

A Staple Since the Dawn of Times

Corn has been a staple in the diets of indigenous people since the dawn of time. Its importance in some indigenous cultures is sacred. Known as maize, almost every tribe grew corn and consumed it at nearly every meal. South American tribes were among the first to cultivate corn and are thought to have created it from wild grasses hybridized over time. Corn wasn't just their main source of food; it was a part of their spiritual practice. Today, corn has evolved into many different species. Some species are better for feeding livestock, some for their sweet taste, and some are even grown for decoration.

Look for heirloom varieties when searching for corn seeds. One of the best heirloom varieties of corn available is called *Golden Bantam*. It is a traditional corn that can be described as sweet, but not too sweet. This variety of corn grows up to six feet tall and will produce anywhere from eight to ten inch ears if it is given the right growing conditions. This variety was first introduced over a century ago. When storing your corn seeds, you can safely have up to two years' worth on hand. Corn seeds may even store longer than that if they are stored properly.

Heirloom varieties of corn are often easier to grow and delicious. You will need a few things to ensure a successful harvest. First, make sure the seeds are planted in well-draining soil. Corn is very picky about the soil and will not thrive in soil that doesn't drain well. Try to plant seeds around two weeks before the last frost. Prepare the soil with plenty of nitrogen by adding manure and blending it into your corn row well. Space your seeds up to 36 inches apart and plant them in a row where they can be exposed to plenty of sunlight. Many heirloom varieties are draught resistant, which may come in handy if you live in an area that doesn't get much rain. Try to plant them in an area of your property where they can be watered if they go too long without rain.

Why is corn so important? For starters, it is a great source of antioxidants, B vitamins (great for energy), and fiber. It is a nourishing crop that sustained indigenous people for centuries. Another reason corn is one of the best crops to grow is that it is versatile. So much can be done with corn besides just eating it off the ear (although boiled ears of corn are delicious). Corn kernels can be dried and processed into flour. Flour is an invaluable staple and can be used to make many other foods like tortillas, corn fritters, batter, and breads.

The corn silk that surrounds the kernels has a valuable use as well. It is diuretic and cleansing to the body. It has been collected, dried, and infused into water to make tea to treat ailments like urinary tract infections and gout.

A Nutrition-Packed Root Vegetable

Carrots are an excellent choice for a hearty root vegetable to add to your survival garden. Carrots come in many varieties, just make sure to purchase heirloom seeds. Carrots are an excellent source of vitamin A, boasting a whopping 428 % of your daily vitamin A value. They are also a rich source of the antioxidant beta-carotene. Beta-carotene is the compound that gives carrots their vibrant color. It is known for its impressive health benefits like preventing cancer. Carrots are also a good source of vitamins C and K, as well as potassium.

In addition to the nutrition factor, carrot seeds are some of the easiest root vegetables to grow. One great thing about carrots is that they tolerate cold weather well. If you are concerned about what crops you may have available in the winter months, you might feel better to know that carrots may still be harvested during this time. Many gardeners actually plant carrot seeds sixty days before the first frost so they can have carrots through the winter. This makes them valuable for year-round nutrition.

Carrot seeds have a moderate shelf life of around two years, but may last longer if stored properly. Remember to let some of your carrots go to seed every year so you can harvest these seeds to add to your stockpile.

Carrots require loose soil without any rocks to inhibit their growth. They do well in a variety of temperatures, ranging from the 40s to the 80s. If you grow carrots in a raised bed, you can grow them just about any time of the year. Make sure you give them plenty of water as they grow. After planting seeds, you can expect to harvest your first carrots in around seventy to eighty days.

The Two Great Grains

Wheat is a classic grain grown for use in a huge range of products from cereals to breads. Wheat has been cultivated since at least 9600 BCE.

It is a rich source of fiber, protein, B vitamins, and minerals like iron, magnesium, copper, selenium, folate, and phosphorus. This alone makes it vital for use in a survival garden.

If wheat seeds are given the right conditions, they can last many years in your seed stockpile. They need to be kept away from any moisture that might cause damage. One of the most effective ways to store wheat seeds is in the freezer. This can extend its shelf life by four to five times.

A variety of heirloom wheat called Emmer wheat (*Triticum turgidum*) is a very good option if you are considering growing wheat.

This is one of the oldest varieties of wheat, often called "pharaoh's wheat" because it was cultivated in ancient Egypt.

This wheat variety is said to be hearty and easy to grow in bad soil. It is also surprisingly disease resistant.

It is great for making anything from bread to beer. If you happen to be one of the millions of people who are sensitive to gluten, this variety of wheat is thought to have less gluten than most others and might be tolerated better than other varieties.

As previously stated, growing Emmer wheat is not as difficult as one might expect. This lies in the fact that it can grow in poor soil and is disease resistant. Start by planting seeds in the spring after the last frost.

They will require well-draining soil, but not soil that dries out too quickly. They do best in full sun. Wait to harvest until there is no longer any milky substance when grains are cut open.

Amaranth is another grain grown since ancient times, having been cultivated for at least 8,000 years.
It is arguably one of the most nutritious crops in the world. It is rich in fiber and protein, which are vital to healthy survival.
However, it also contains a large amount of micro nutrients that are vital to health, such as iron, magnesium, selenium, phosphorous, and copper. It is packed with antioxidants that help to destroy disease-causing free radicals in the body.
This super food has been studied for its ability to lower inflammation in the body, reduce cholesterol that could lead to heart disease, and even aid in weight loss.
It does not contain any gluten, making it even more beneficial for those sensitive or unable to have any gluten in their diet.
Amaranth grows best in soil rich in nitrogen and phosphorous, but will still grow well (up to six feet tall) in the average garden soil. If given the right soil combination, amaranth plants can grow very tall, up to eight feet in height.
Try planting amaranth seeds in well-draining loam. They will need at least five hours of sunlight each day to thrive.
Although they seem to tolerate somewhat drier soils, they do better with regular watering. Do not plant seeds until danger of frost has passed.
Once amaranth starts to grow, you can use it in two different ways. After fifty days, you can use the nutritious greens in salads.
After ninety to one hundred days, you can begin to harvest its nutrient-dense grains to use.
Amaranth grains can be processed into flour for baking or added to virtually any meal as a valuable source of sustenance and nourishment.

The Grandiose Greens

Green may not be your favorite color, but in the vegetable world, green is synonymous with "great".
Plants like spinach, peas, and asparagus are nutritional powerhouses that virtually all contain beta-carotene, vitamin A, vitamin C, vitamin K, magnesium, potassium, and iron, to name a few. In addition to these benefits, all three of these plants are relatively easy to grow.

Spinach is a leafy green that is a favorite among nutritionists around the world. This is because for every three cups of spinach you consume, you are getting roughly 300% of your daily vitamin K intake, 160% daily vitamin A, and 40% of the vitamin C you need each day.
You are also getting at least 45% of your daily intake of folate. It is a good source of antioxidants and has been studied for its ability to support brain health, lower blood pressure, and support healthy eyes.
Spinach actually grows better in cooler weather, so you can plant in early spring and then again in the fall. This way you will have spinach to enjoy twice a year. Sow the seeds in thin rows and try to space rows at least twelve inches apart. Lightly cover the seeds with soil. Plant seeds in a location that has plenty of shade, because spinach does not do well in the full sun.
Keep the soil moist until the leaves germinate, then water as needed. Once you see leaves begin to emerge, try to thin out your rows to keep plants six inches apart. Harvest all of your spinach leaves before they go to seed. After they go to seed, the leaves are too bitter. Let a few go to seed so you

can harvest seeds to stockpile. Spinach seeds will keep up to four years, and perhaps longer if stored properly.

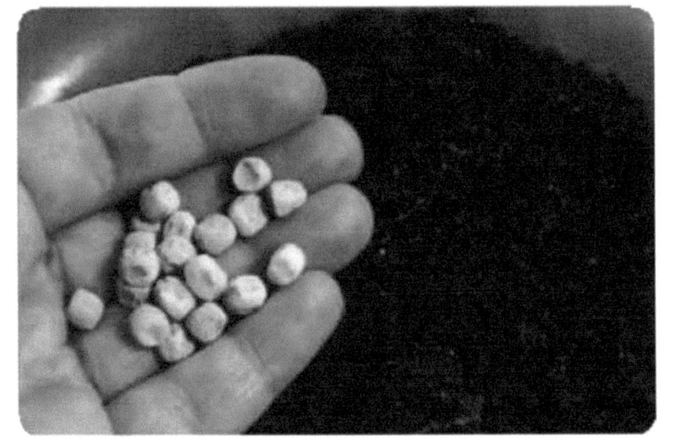

For a crop rich in fiber, protein, and plenty of vitamins and minerals, stock up on peas. Peas can be an important source of plant-based protein, making them valuable for survival gardens.

Their fiber content makes them a wise choice for gut and digestive health.

Another benefit of this crop is the shelf life of the seed, which is around three years. However, they may last longer than this if they are stored properly.

The simplicity of growing peas is another reason why they are a valuable seed to have in stock.

Peas can be planted as soon as the ground is workable. They require a growing temperature between 55 degrees and 65 degrees Fahrenheit for the most success. They prefer soil that is loamy and has a pH between five and seven.

Create a shallow trench around 22 centimeters wide and set seeds down inside. When placing seeds in the trench, place them down 3 centimeters deep and space them out around ten centimeters apart.

They will do well in full sun and will require a trellis to climb, as they are vining plants.

Asparagus is the gift that keeps on giving in the vegetable world. This is because it is a perennial vegetable that can produce crops for twenty years or more!

Additionally, asparagus is packed with nutrition and health benefits.

This useful crop has been studied for its ability to lower blood pressure, reduce allergy symptoms, and even protect against some diseases.

It is a mild diuretic, so it has been used traditionally to help flush out toxins from the body and promote a healthy urinary tract.

Asparagus can be planted in the spring after any danger of frost has passed. Start by digging a trench around six to twelve inches wide and six to twelve inches deep.

Choose an area of your garden that gets plenty of sun and has well-draining soil. Space seeds around twelve inches apart.

Lightly cover them with dirt. Water daily until germination, but avoid over watering.

Once these plants start growing, they are not as picky about their growing conditions.

Plant plenty of seeds your first go-around, because they take time to grow enough to harvest for food.

It generally takes about four years before you will have good-eating asparagus to harvest.

If you planted a lot of seeds the first year, you will have many plants to harvest in the following years since these plants come back yearly and last for many years.

The seeds have a relatively good shelf life of around three to four years.

A good garden is worth the time and effort you put into it.

With the seven seeds discussed in this article you can create a lot from a little and support yourself, as well as your family, through any crisis.

With the right seed storage and cultivation knowledge, you will be well on your way to having the ultimate survival garden setup with optimal sus-